QUEERING BATHROOMS: GENDER, SEXUALITY, AND THE HYGIENIC IMAGINATION

The gendered nature of public washrooms has become a source of social anxiety and political controversy in recent years: queer and trans folk have been harassed for allegedly using the 'wrong' washroom, while activists have campaigned for more gender-neutral facilities. In *Queering Bathrooms*, Sheila L. Cavanagh explores how public toilets demarcate the masculine and the feminine, and condition ideas of gender and sexuality.

Based on one hundred interviews with gay, lesbian, bisexual, transgendered, and/or intersex peoples in major North American cities, the book delves into the ways that queer and trans communities challenge the rigid gendering and heteronormative composition of public washrooms. Cavanagh incorporates theories from queer studies, trans studies, psychoanalysis, and the work of Michel Foucault, and argues that the cultural politics of excretion is intimately related to the regulation of gender and sexuality. Public toilets house the illicit and act as repositories for the social unconscious. Offering suggestions for imagining a more inclusive public washroom; *Queering Bathrooms* asserts that although the subject of toilets is not typically considered within traditional scholarly bounds, it forms a crucial part of our modern understanding of sex and gender.

SHEILA L. CAVANAGH is an associate professor in the Department of Sociology at York University.

SHEILA L. CAVANAGH

Queering Bathrooms

Gender, Sexuality, and
the Hygienic Imagination

UNIVERSITY OF TORONTO PRESS
Toronto Buffalo London

©University of Toronto Press Incorporated 2010
Toronto Buffalo London
www.utppublishing.com
Printed in Canada

ISBN 978-1-4426-4154-9
ISBN 978-1-4426-1073-6

Printed on acid-free, 100% post-consumer recycled paper with vegetable-based inks.

Library and Archives Canada Cataloguing in Publication

Cavanagh, Sheila L. (Sheila Lynn), 1969–
Queering bathrooms : gender, sexuality, and the hygienic imagination/
Sheila L. Cavanagh.

Includes bibliographical references and index.
ISBN 978-1-4426-4154-9 (bound).   ISBN 978-1-4426-1073-6 (pbk.)

1. Public toilets – Sex differences.   2. Public toilets – Social aspects.
3. Gender identity.   I. Title.

GT476.C39 2010   628.4'508   C2010-904493-2

This book has been published with the help of a grant from the Canadian Federation for the Humanities and Social Sciences, through the Aid to Scholarly Publications Program, using funds provided by the Social Sciences and Humanities Research Council of Canada.

University of Toronto Press acknowledges the financial assistance to its publishing program of the Canada Council for the Arts and the Ontario Arts Council

 Canada Council   Conseil des Arts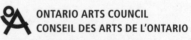
for the Arts   du Canada

University of Toronto Press acknowledges the financial support for its publishing activities of the Government of Canada through the Canada Book Fund.

Contents

Acknowledgments vii

Introduction 3

1 Queering Bathrooms: Gender, Sexuality, and Excretion 27

2 Trans Subjects and Gender Misreadings in the Toilet 53

3 Seeing Gender: Panopticism and the Mirrorical Return 79

4 Hearing Gender: Acoustic Mirrors – Vocal and Urinary Dis/Symmetries 105

5 Touching Gender: Abjection and the Hygienic Imagination 134

6 Sexing Gender: The Homoerotics of the Water Closet 169

Conclusion 205

Glossary, by Sheila L. Cavanagh and Melissa White 223

Notes 227

Bibliography 259

Index 275

Acknowledgments

I began writing this book in 2007 while on sabbatical from York University. This book would not have been written without the support and academic feedback of many friends and colleagues. I spent my sabbatical in Brisbane, Australia, where I had a visiting professorship in the Centre for Critical and Cultural Studies (CCCS) at the University of Queensland (UQ). I am grateful to Graeme Turner, director of the CCCS, for finding me a quiet place to write in the centre during my stay. I am equally thankful to Peter Cryle, director of the Centre for the History of European Discourses (CHED), for his generous invitations to dinner where I met many great scholars affiliated with the research centre. I also want to give a special thanks to Elizabeth Stephens and Alison Moore for their friendship and great conversations over wine in Fortitude Valley. Alison Moore's scholarship on the history of excretion and psychoanalysis greatly influenced the trajectory of my work. I am especially grateful to her for her contributions to my theoretical analysis. I am most indebted to Louise McCuaig, who gave me a room to work in her beautiful home while I was on sabbatical. Her friendship and generosity were unparalleled. I cannot thank her enough for everything she did to make my time in Australia memorable and for introducing me to the academic community at UQ.

Upon my return to Toronto, I was fortunate to have many critical eyes on the manuscript during the review process. I wish to thank the anonymous reviewers at the University of Toronto Press (UTP), who enthusiastically supported the manuscript. I thank Virgil Duff, executive editor of UTP, for his praise and untiring support. Anne Laughlin, managing editor, skilfully supervised the copy-editing process. She generously gave me extra time and indulged my need to make changes to the manuscript late in the production process. Douglas Hildebrand was particularly flexible, wise (and diplomatic) during our negotiations on the book title and cover page. Jenna Germaine, catalogue and copy coordinator, was responsible for getting book endorsements from the wonderfully accomplished and important scholars quoted on the back cover of the book, and for guiding me in the editing of the book description (now published in the UTP catalogue). I also wish to acknowledge the Aid to Scholarly Publications Programme for a grant in aid of publication and the generous support of the Social Sciences and Humanities Research Council of Canada, which made this book possible through Standard Research Grant number 410-2008-2022.

I owe a tremendous debt of gratitude to a cluster of brilliant graduate students at York University whom I employed as research assistants to conduct interviews for this book. Without their ingenuity and untiring commitment, the research project would not have been possible. I give special thanks to Erin Bentley, Diana Gibaldi, Vicki Hallett, Lee Wing Hin, Beenash Jafri, Habiba Nosheen, Marc Sinclair, and Rob Teixeira for the interviews they completed, along with their contributions to the implementation and design of the study. I also wish to give my profound thanks to Caleb Nault and Laine Hughes for reading an earlier draft of this manuscript and offering invaluable criticism. Melissa White did an excellent job as co-author of the appendix and in finalizing the bibliographical notes.

I also want to recognize the SexGen committee at York University for the important work they have done on the gender-neutral bathroom campaign; the Sexuality Studies Council at York University for promoting gender and sexuality studies at the university and in the LGBTI community more broadly; the Sociology Department and the Women's Studies Department at York for supporting me in the publication of this study; and finally, Kyle Scanlon, the trans program coordinator at the 519 Church Street Community Centre in Toronto for his wonderfully helpful advice and guidance with respect to the *Queering Bathrooms* questionnaire and for his work on public bathroom accessibility throughout the city.

I want to give a special thanks to all of the interviewees who trusted me with their stories. I am grateful for their time and their willingness to participate in the *Queering Bathrooms* project. Without their generosity this book would never have been written.

I also wish to thank friends and colleagues who, in various ways, supported me in seeing this book through to publication: Kym Bird,

Alison Crosby, Emma Donoghue, Andil Gosine, Robert Grant, Judith Hamilton, Frances Latchford, Radhika Mongia, David Murray, Eric Mykhalovskiy, Bobby Noble, Chantal Phillips, Chris Roulston, Marc Stein, Arja Vanio-Matilla, and Leah Vosko. While some of you offered personal insights about the cultural politics of bathrooms, and others offered friendship and/or academic support in numerous other ways, I am sincerely grateful to you all. Finally, I wish to thank Donia Mounsef, who read the manuscript (almost) in its entirety and offered deft theoretical criticism and encouraging words.

QUEERING BATHROOMS:
GENDER, SEXUALITY, AND
THE HYGIENIC IMAGINATION

Introduction

Bathrooms ... accentuate otherness.

(Cummings 2000, 271)

Under the spell of the modern goddess Hygiene, bathroom design has lost its capacity to become the focus of relaxing psychic activity. Instead, the bathroom has become a place – or better, a metonymic space – a closet – of secret constraints.

(Frascari 1997, 165)

The turn of the twenty-first century is a key moment in the cultural politics of gender and excretion. Those who are not transgender seem to be uncertain about the relationship between the body and gender identity. In her discussion of *Fin de siècle, Fin du sexe* (an epigram used by French artist Jean Lorrain to signal late-nineteenth-century angst about gender confusion), Rita Felski (2006) notes that the trope of transgender figures prominently in present-day worries about the status of sex. This worry is well evidenced in the modern lavatory. Even a cursory glance at the international media confirms that there is an international preoccupation with public bathrooms, gender identities, and rights of access. Consider the following international news stories: 'Transsexual teen's use of girls' toilets raises fears: Parents fight B.C. [Canada] school's decision'; 'Trans activists charge harassment in New York City: Police selectively requiring ID at public bathrooms at Christopher St. riverfront'; 'Transgender MP in toilet fracas: An Italian opposition MP and former showgirl has expressed outrage after meeting a transgender colleague in the Parliament's ladies' toilet'; 'Thai school adds "third sex" bathroom for transgender students'; 'Transvestie water closets [in Rio De Janeiro]'; 'De-gendered loos spark row in UK'; 'Patron lodges complaint over toilet row in pub in [Cairns, Australia] sex bias claim'; 'Universities heed the call for genderless washrooms [Canada].'

As illustrated in these headlines, transgender is a point of controversy and debate. Anxiety about gender variance is keenly felt in public washrooms and often projected onto trans folk. People worry about what Michel Foucault (1978) calls a truth about sex, and nowhere are the signifiers of gender more painfully acute and subject to surveillance than in sex-segregated washrooms. Transgender denotes the 'dissolution of once stable polarities of male and female, the transfiguration of sexual nature into the artifice of those who play with the sartorial, morphological, or gestural signs of sex' (Felski 2006, 566). If the zeitgeist of the late nineteenth century is characterized by a mixture of despair and opulence, degeneration wrought by a dangerous excess and impending doom – part of which is signified by tropes and images of gender disorder – then it is fruitful to reflect upon the fin de siècle of the late twentieth century. The perceived loss of a binary gender axis, the grid upon which normative heterosexuality depends for cogency and intelligibility, incites anxiety about gender incoherence. Those who are ill at ease with transgender and transsexual, lesbian, gay, bisexual, and/or intersex people are, ironically, posing questions about their own safety, rights to privacy, and access to public washroom facilities in ways that are unique to the present. While there is a complete lack of evidence to substantiate any actual infringement on the civil liberties of people who are conventionally gendered and cissexual (non-trans) (Serano 2007), such persons are, nevertheless, territorial and defensive about the gendered composition of the toilet. As I will argue in the following chapters, there is nothing rational or legitimate about gender panic in modern facilities. The upset is irrational and about what Richard Juang (2006) calls 'heterosexual genital security' (716).

The management of the body and its modes of evacuation in the modern lavatory are part and parcel of the regulation of gender and sexuality in the modern era. The elimination function is an area of bio-political regulation that is often designated 'out of scholarly bounds' (not to mention crude and subject to interdiction in polite discourse), yet, curiously, central to an overemphasis upon an absolute and unchanging sexual difference. Robyn Longhurst (2001), in her study of bodies, fluids, and excremental space, notes that toilets are 'one of geography's abject and illegitimate sites that have been deemed (perhaps unconsciously) inappropriate and improper by the hegemons in the discipline' (131). Yet ideas about what is a 'proper,' 'worthy,' and 'respectable' topic of inquiry operate to censor, repress, and prohibit but also, as Foucault (1978) tells us, to map the terrain of the thinkable, the analysable.[1] *Queering Bathrooms* argues that obsessive investments in 'urinary segregation' (Lacan 2006) by gender is about a perceived threat to sexual difference and to heteronormativity. 'It becomes considerably harder to delineate who is gay and who is lesbian when it is not clear who is male or a man and who is a female or a woman' (Devor and Matte 2006, 387).

This book endeavours to theorize *how* and *why* the public washroom is a site for gender-based hostility, anxiety, fear, desire, and unease in the present day as the washroom is also a site of homoerotic desire. *Queering Bathrooms* is concerned with the nuances and vicissitudes of trans and homophobic hate and the harm incurred by gender-exclusionary spatial designs. My analysis seeks to expose how sex-segregated designs function to discipline ways of being gendered that are at odds with a normative body politic. I build upon questions posed by Peg Rawes (2007) in her discussion of Luce Irigaray and sexed architecture: 'What subjects are considered to be too complex for architectural design, and are therefore rejected as excessive or unnecessary to its aims? How successful is architecture at developing spaces for specific users, if it does not take into account the needs of the sexed subject?' (77). While Irigaray wants to make room for the feminine or for the 'sexed subject' (who is usually coded as 'woman' in her analytic), she is also, in my reading, asking how we can imagine an architecture that does not reproduce a universal humanist (or phallocentric) sameness beholden to a sanitized and unsexed ideal. The institutionalization of gender-neutral toilet designs is an urgent and important political project to ensure access for all who depart from conventional sex/gender body politics. But it is equally important to think creatively about how we may build gendered architectures that prompt people to think about gender, sexed embodiment, desire, and our relations to others in new and ethical ways.

Using interviews with those who are lesbian, gay, bisexual, transgender, and/or intersex (LGBTI) in Canada and the United States, I first argue that the institution of the public toilet is designed to discipline gender. As I discuss in the chapters that follow, gender is disciplined through the gendered codes of conduct and the hygienic and panoptic designs of the modern lavatory. Michel Foucault emphasizes the use of panopticism to produce what he called the transparent society. Bodies were made visible in the eighteenth century so that they could be subject to surveillance. Foucault highlights the use of modern optics to show how the supervisory gaze was concerned not just with disciplinary power but with self-government. The minutiae of our everyday lives are now subject to visual inspection. Panopticism prioritizes vision, and there has been a concurrent degradation of the remaining sensory systems from the eighteenth century through to the present day. In fact, the 'discipline of policing and sanitation depended in turn upon a transformation of the senses' (Stallybrass and White 2007, 273). This transformation is well evidenced in the history and cultural politics of the European and British water closet. The bodies of the poor and diseased were subject to a panoptic gaze. Touching and smelling and, to a lesser extent, hearing had to be subordinated to a pure and unmitigated hygienic gaze: the 'policeman and soap are analogous' (ibid., 272). Focusing on modern plumbing in private and public domains, Nadir Lahiji and D.S. Friedman (1997) also consider how the supervisory gaze internalized by the subject is hygienic. A 'hygienic superego' (53) takes shape to quell pleasures taken in unclean, racialized, and gender-variant bodies. Gender, like hygiene, was to be made visible so that the subject could learn to govern him or herself. This form of government operates upon individual psyches. In the Foucauldian economy of discipline the 'supervisory gaze is also hygienic' (ibid.). 'Cleanliness is the response to a guilt modernity has had to internalize ... The superego of the hygienic movement constructs modernity by plumbing the destructive instinct of the pleasure we take in dirt and pollution' (ibid.). Trans and/or queer bodies – not infrequently read as dirty, polluted, or unintelligible (albeit in different ways) – are subject to a persecutory gaze in the modern lavatory by those who are transphobic and/or homophobic. Trans bodies are already subject to more formalized systems of bio-political regulation in medical, scientific, and legal discourses and institutions, as the literature on sex reassignment makes clear (Stryker and Whittle 2006). Public restroom facilities stage gender so that non-experts – the general public – can decide if it is pure and intelligible or impure and indecipherable in heteronormative and cissexist landscapes.[2]

Using queer theory, trans studies, psychoanalysis, and the seminal work of Michel Foucault (1979) on the disciplines, I show how modern lavatories are gendered and shaped by a white hygienic superego. There is a metonymic relationship between gender variance, danger, dirt, and disease in the present-day restroom. As in the late nineteenth century, when 'police and soap ... were the antithesis of the crime and disease which supposedly lurked in the slums, prowling out at night to the suburbs' (Stallybrass and White 2007, 273), the contemporary toilet is a place where gender variance and homosexuality are linked to dirt, disease, and public danger. Those who are recognizably trans are subject to persecution for using the 'wrong bathroom' in ways that are not only callous and cruel but compulsive and curious. The urgency with which one seeks to clarify the gender identity of another or to expunge gender-variant folk from the public lavatory entirely is beset by worries about disease and disorder that, in the present day, are overlaid by angst about a racialized and class-specific gender purity.

It should be remembered that the institution of the gender-segregated toilet, now commonplace in Canada and the United States, dates back to nineteenth-century England and France. The development of the modern water closet along with the building of city-wide sewer systems in London and Paris; new scientific understandings of germ theory and disease-control methods (specifically, for cholera, smallpox, and typhus); and changing moral codes governing personal hygiene and sanitation – all imported to Canada and the United States (and imposed upon European colonies) – are key to situating what Alexander Kira (1976) calls the 'sex-elimination linkage' (107). Kira, associate professor of architecture at Cornell University in New York, noted in 1966 that discomfort with and shame about the sex organs affect the design and usage of public toilets. 'Because of a fairly direct anatomical and neuromuscular correspondence between the body parts used for elimination and those used for sex, our attitudes toward sex are also linked to our attitudes toward elimination' (1966, 54). He even suggested that concern about urinary noise-making in public facilities (to be discussed chapter 4) is driven by how the noise sounds like the act of sex. Kira contends that the link between sex and elimination is rooted in 'Anglo-Saxon insistence on privacy on a sexual basis: the existence of men's and women's rooms, which guarantee complete privacy from the opposite sex but only limited privacy from members of the same sex' (107). The focus on modesty and privacy in toiletry habits strongly resembles protocols for hetero-sex. The heteronormativity governing the gender of urinary designs is so deeply entrenched in dominant North American cultures that it often escapes notice.

The gender-segregated facilities of concern in this book reproduce what Judith Butler (1990) calls the heterosexual matrix. While trans theorists have developed important critiques of Butler's theory of gender performativity (Namaste 2000; Prosser 1998, 2006) and of the way queer theory more generally appropriates trans bodies in its framing of gender fluidity (Boyd 2006), I find Butler's analytic of the heterosexual matrix useful in thinking about what anchors gender to thought about sexuality in the normative landscape. Butler defines the heterosexual matrix as a 'grid of cultural intelligibility through which bodies, genders, and desires are naturalized' (151). Gender and sexuality are, in her paradigm, mutually constitutive and reinforcing. Heteronormativity is dependent upon an absolute split between male and female gender identities, their circulation, repetition, and recitation in the social milieu. Sexualities consolidate but can also undo and trouble gender. Bodies are interpolated by masculine or feminine modes of address in what Julia Serano (2007) calls cissexist culture. Cissexist cultures (as I will discuss more fully in chapter 2) are predicated upon the idea that transsexual identities are inauthentic or inferior copies of those gender identities had by non-trans folk. Butler's analytic of the heterosexual matrix enables us to consider how heteronormativity is dependent upon a means of sorting bodies into two divergent and mutually exclusive gender locales. Male is to masculinity as female is to femininity, and the normative concord underpins the institution of heterosexuality.

The heterosexual matrix is, in part, sustained by gender-exclusionary spatial designs and the ways we are imposed upon to navigate those spaces. For those whose gender is not legible to non-trans folk, or for those whose sex is read as being at odds with gender signs on bathroom doors, public space accessibility and safety are at stake. The daily news is full of stories about gender troubles, toilets, and rights of access denied.[3] Consider the following examples: In New York the Hispanic AIDS Forum won their suit against the landlord for refusing to 'renew the lease on their Jackson Heights office because of complaints from other tenants that transgendered clients were using the "wrong" restrooms' ('Judge's rulings favor Latino AIDS agency in transgender discrimination suit,' 16 January 2003). Arpollo Vicks, an African-American Hurricane Katrina evacuee, was jailed (and released five days later) for showering in the 'women's' washroom at a shelter in Texas (Hensley, 'Transgendered evacuee released,' 10 September 2005).[4] Complainants believed she posed a threat to non-trans women and children in the semi-public shower.[5] Controversy ensued when Kampang Secondary School, in the north-east of Thailand, built a transvestite toilet, signposted 'half-man, half-woman ... in blue and red' (Jonathan Head, 'Thai school offers transsexual toilet,' BBC News, 29 July 2008) on the door.

In Canada, Leslie Morgan of Tilbury, Ontario, resigned from her job at the Siemens VDO plant following intensive transphobic harassment that involved the washroom she used being 'fouled up by urine – on the toilet tank, seat and surrounding area' ('Threats cost me job: transsexual,' *The Gazette*, 29 March 2005).[6] Messages were also scrawled on the bathroom wall (Morgan used her own separate bathroom), including: 'No Freaks Here' and 'Freaks Go Home' (ibid.). Morgan filed a complaint with the Ontario Human Rights Commission (Pearson, 'Harassment draws fire,' *Windsor Star*, 30 March 2005). In British Columbia, Leslie Ferris, a transgender woman, filed a complaint (which she eventually won) under section 8 of the Human Rights Act in 1998 alleging that Beach Place Ventures Limited and the Office and Technical Employees Union, Local 15, had discriminated against her. The complaint was based on the company's refusal to allow Ferris to use the 'women's' washroom following a complaint made by another worker about a 'man using the women's washroom.'[7]

Of course, one only has to be perceived to be transgender and/or gay to face physical assault in public bathrooms. Willie Houston of Nashville, Tennessee, is a case in point. He was a 'straight man using a men's restroom under an unfortunate combination of strange coincidences featuring a purse, a friend and a homophobe ... holding his fiancée's purse in one arm and guiding a blind male friend with the other ... Another man hurled anti-gay insults at Houston, and the situation escalated until Houston was shot once in the chest. Among his last words to his fiancée were, "Just remember, I will always love you"' (Morgan, 'Reading the fine print on the bathroom door,' *City Pulse*, 6 December 2006).

In another disturbing case marked by violence, seventeen-year-old Gwen Araujo was beaten and strangled to death in a private bathroom in San Quentin by three young men in October 2002. Two of the men involved had oral and anal sex with Araujo and later 'discovered her gender after cornering her in a bathroom during a party at Merel's [one of the attackers] house in Newark' (Lee, 'Nabors apologizes for role in Araujo slaying,' *San Francisco Chronicle*, 25 August 2006). During the attack one assailant cried out 'I can't be f– gay' (ibid.). Evidently, Araujo's transgender status upset the attacker's certainty about his own sexual identity.

In Canada, Alexis McKay, a trans woman, was also physically assaulted by Randall Allan Viggers after he met her in the unisex washroom at the Woodbine Hotel in Ontario. McKay agreed to give Viggers oral sex (for money), but once McKay began to question her gender identity (he saw what he perceived to be male genitals), he beat her 'in the face, with a beer bottle and a table leg ... she required 150 stitches, mostly to repair slashes and cuts to her face' ('Brockville man jailed,' *Kingston Whig-Standard*, 26 September 2002).

The anxiety and aggression surrounding gender are, as shown in the above-mentioned cases of physical assault, very much about optics, aggressive disidentifications (I am not a 'fag' or a 'tranny-queer'),[8] attempts to police the borders governing the self (don't touch me or my 'girlfriend'), and heteronormativity. The violence and policing of gender in toilets is an indicator of the lengths to which non-trans people will go to defend their allegiances to allegedly natural and mutually exclusive gender positions. There is, as Lee Edelman (1993) suggests, a uniquely non-trans, white, heteromasculine anxiety about sodomy (ass fucking), and the breaking down (real or perceived) of the dominant signifiers of sexual difference – such as the inability of the signs 'Ladies' and 'Gentlemen' on the toilet door to effectively regulate who goes into which room. Of course, non-trans women can also be intolerant of trans people in the toilet. I rework dominant heterosexist and transphobic safety narratives in chapter 2 to account for a specifically cissexual female form of disdain and intolerance. The anxiety focused upon LGBTI people often manifests itself in aggressive disidentifications in ways that are in some moments homophobic and in other moments transphobic. Sometimes the two forms of intolerance converge. I do my best in this book to recognize the distinctions between transphobia and homophobia and to identify the moments when the two modes of intolerance intersect and are mutually constitutive.

In Canadian and American public restroom facilities, people tend to regulate gender and interpersonal encounters with a defensive exuberance.[9] The atmosphere in these facilities is structured by an obsessive interest in gender integrity and interpersonal boundaries. There is a will to secure the signifiers and bio-politics of sexual difference. The feminine body is supposed to be kept clean and neat, bound and private. Masculine bodies are, by contrast, authorized to make more fecal and urinary sounds; spills and smells are amplified to establish a wider circumference around the body and to stake a territorial claim. While the masculine tends to be well defined and ultra-visible, the feminine tends to be ill-defined and enclosed by partitions. 'At the level of the symbolic, the feminine is said to be on the side of the abject, the irrational, the unformed, the horizontal, the liquid, like bodies of water that take the form of their vessel; just as the masculine is said to be on the side of the subject, the rational, the normative, the distinct, the vertical, the categorical, the specific' (Morgan 2002, 172).

The oval pedestal enclosed by stall partitions functions to quarantine (and ensure) the purity of a white, feminine subject position, while the masculine is displayed before the urinal in full-frontal (open) view. Studies of public toilet designs institutionalized in Canada and the United States sometimes note that the urinal resembles a vagina and the enclosed oval toilet bowl resembles an anus (Kira 1966, 1976). The former is public and visible whereas the latter is hidden by stall partitioning and a closable door. Receptacle shapes and designs, along with the positioning in lavatory space, seem to mirror human anatomy. There are, of course, significant cultural and class-specific variations to the way these ordinances are enforced – depending upon the context and location – but the generic design is meant to accentuate the difference between feminine and masculine urinary positions, male and female. The gendering of the modern, porcelain toilet oval as white is not to go unnoticed. Critical geography has, in recent years, been concerned with the spatial production of whiteness (Dwyer and Jones 2000; Nast 1998, 2000), and postcolonial theorists have noted the centrality of the English water closet to processes of racialization by European and American colonial and imperial powers (Anderson 1995, 2002; Anspaugh 1995; Chun 2002; Cummings 2000; Inglis 2002; Largey and Watson 1972; Morgan 2002; Srinivas 2002; Van Der Geest 2002). 'Abby Rockefeller, who manufactures composting toilets, wrote in her preface to Sabbath and Hall's book *End Product*, "As the missionary bibles once spread light to savage hearths around the globe, the porcelain john will soon spread the Word of Technology to every home of the Great Unwashed"' (Lewin 1999, 69).

Critical geographer Steve Pile tells us that 'The body-ego-space is territorialized, deterritorialised and reterritorialised – by modalities of identification, by psychic defence mechanisms, by internalized authorities, by intense feelings, by flows of power and meaning' (1996, 209). Sexual difference is produced, in part, through a culturally and class-specific spatialization of sensory systems. We are pinned to body maps and spatial grids by gender. Wafting smells, peculiar sounds, queer sights, and body residue prompt interpersonal angst and upset. Teresa Brennan (2004) observes that the 'Western individual [is] especially more concerned with securing a private fortress, personal boundaries, against the unsolicited emotional intrusions of the other. The fear of being "taken over" is certainly in the air ... Boundaries, paradoxically, are an issue in a period where the transmission of affect is denied' (15). People are concerned about wafting smells, body-echoes, unusual sights, and illicit touches because these are felt to interfere with gender and subject integrity. Enclosed lavatories are sometimes experienced as extensions of the subject, psychically invested bodily prostheses.

Space is therefore relevant to the way we negotiate gender identity and social difference. Geographers of urban space have been turning to psychoanalysis to interpret what Steve Pile (1996) calls the 'psychodynamics of space' (Aitken 2001; Aitken and Herman 1997; Callard 2003; Kirby 1996; Lefebvre 1998; Nast 1998; Philo and Parr 2003; Rose 1993, 1996; Sibley 1995, 2003; Wilton 1998). Building upon analysis of urban space done by David Sibley (1992) and Gillian Rose (1996), along with conceptions of dirt as 'matter out of place' (Douglas 1966), Steve Pile insists that we must understand how vital social organizations of space are to the alignment of socially disenfranchised people with pollution (he uses the example of Gypsies in Britain), branding them 'foreign,' 'outsiders,' 'threatening,' and in various ways 'other' to the social body.[10] In his discussion of space, intolerance, and difference, Wilton (1998) argues that a link between the social, the psychic, and the spatial is necessary to understand the anxiety emerging when people encounter 'difference' (socially disenfranchised others) and feel their perceptions of self (and others), along with their body boundaries, to be unstable. This is true in the arena of gender and in other identitarian registers used to secure bodily egos (psychically invested fantasies of the body's surface). David Sibley (2003) also concurs that 'anxieties about others, and the regulation of space are central to debates about socio-spatial relations' (395). These formulations underscore the importance of public space to the production of gender and social difference more generally.

In his discussion of urban space and subjectivity, Pile further comments on the geographical and psychic borders set in place to police and contain demonized others. 'The wall is an armour against the other – it is meant to shore up those feelings of "fear" and "disgust": a hard, high, fixed, impermeable boundary on a space which is both urban and bodily' (Pile 1996, 6). Social space is designed to render people 'in' and 'out' of place. LGBTI folk are often rendered 'out of place' in public lavatories. This is because gender and sexual minorities are perceived to disrupt the cissexist and heteronormative designs of the bathroom. Those who are recognizably trans are seen to be 'out of place' because the cissexist cultural landscape demands a concord between gender identity and sexed embodiment. It must also be said that gays and lesbians who are not trans but invested in what Lisa Duggan (2003) calls 'homonormativity' are sometimes critical and intolerant of trans people in public space. Even when a given space is designated gay positive it is not necessarily trans positive and accessible. Gender policing in toilets at gay bars and at other allegedly queer-positive community events is a case in point.[11]

Gender policing in toilets occurs because people depend upon the rooms to secure identificatory coordinates. My interpretation of the interview testimonials suggests that people who are not open to questioning and interrogating cissexist orderings of space defensively project their own disavowed desires and identifications onto those who are trans, or onto those perceived to be in defiance of conventional sex/gender systems. Aggressive disidentifications are projective. 'A projection is what I disown in myself and see in you; a projective identification is what I succeed in having you experience in yourself, although it comes from me in the first place' (Brennan 2004, 29). In other words, those who are not trans project their own difficulties with gender and subject integrity on to trans people in negative ways. Closely related to abjection (which is to expel or cast away), the concept of projective identification coined by Melanie Klein in 1946 involves the 'phantasy that it is possible to split off a part of one's personality and put it into another person' (Salzberger-Wittenberg 1970, 138). Conversely, people may also construct others as similar to the self; thereby eradicating the differences residing in the other. Identifications are thus incorporative and/or excorporative. Indicators of such identifications are mapped onto space. 'For, as psychoanalytically inclined geographers have elegantly demonstrated, borders are places of often frantic and violent boundary making – and such boundaries can be strengthened both by incorporating the other (and hence turning its difference into sameness) and by expelling the other (thereby rejecting it as unacceptably alien)' (Callard 2003, 297). The architecture and design of sex-segregated bathrooms accentuate perceived similarities and differences between people in the domain of gender in ways that are 'race' and class specific.

Methodological Notes

Queering Bathrooms is based on a theoretical engagement with analysis and discussion of 100 in-depth interviews with LGBTI people. Because identity categories – their meanings, modes of expression, and psychically invested corporeal and erotic coordinates – are historically and culturally variable, and subject to contestation in today's LGBTI communities, a note on definitions is in order. In Canada and the United States there are significant differences between those who identify as 'gay,' 'lesbian,' 'bisexual,' 'transgender,' 'transsexual,' and/or 'genderqueer,' and also among those who are intersex.[12] While I do not assume a truth about any given gender identity or sexed position, I take care to recognize people's stated identities when referencing interview data. I list participant gender and sexual identities *exactly* as they write and describe them on the demographic information sheet. In some instances, a gender identity but not a sexual identity is given. The reverse is also sometimes the case. This accounts for why I sometimes list participants' sexual identifications and not their gender identifications, and vice versa. Some interviewees identify as 'queer,' and this signifies their gender *and* sexual identities. While the variable usage of the term 'queer' may seem confusing to readers who may be accustomed to thinking about 'queer' as a non-normative sexual orientation as opposed to a gender identity, I choose to respect interviewee self-definitions without modification or qualification. David Valentine (2007) notes in his discussion about the modern cultural politics of naming research subjects that there is a danger in enforcing absolute distinctions between gender and sexuality – particularly when subjects do not abide by them – and that researchers should avoid imposing an unequivocal 'modernist telos wherein the recognition of gendered and sexual identification as separate (if related, in some unspecified way) is more accurate, more true, more valid' (245). A significant number of participants identify as 'genderqueer' in this study, and this designation is often meant to designate a sexual orientation as well as a gender identity. While there is no one-to-one correspondence between a stated gender or sexual identity and one's experience of that identity (including the meanings one attributes to a given identification), I provide an appendix at the end of this book listing and provisionally defining the identities used by interviewees. I take it as given that those who affiliate themselves with a given label may have vastly different gender embodiments, sexual orientations, and practices. My appendix is meant as a guide or road-map to thinking about diverse gender identities rather than as a final or transcendental statement about such identities.

One's experience of gendered space is often determined by the degree to which one is recognized by others as the gender to which one identifies. How one presents the self in the arena of gender, the extent to which the presentation is validated by cissexual onlookers, and the ease with which one achieves a naturalized performance of the gender to which one holds allegiance have an enormous impact on how one experiences the gendered composition of the toilet. 'Bodies that inhabit or enact naturalized states of being remain culturally intelligible, socially valuable, and as a result, gain and retain the privilege of citizenship and its associated rights and protections' (Boyd 2006, 421). It is abundantly clear that those who are sexual minorities (lesbian, gay, bisexual, etc.) do not necessarily have difficulty gaining access to gendered toilets unless they are also trans or perceived to be gender-variant (butch women, effeminate men, etc.) or queer. While non-trans lesbian, gay, and bisexual interviewees do sometimes experience homophobia and gender-based discrimination in public restroom facilities, they do not always have access issues comparable to those of people who are trans or perceived to be gender variant. Trans interviewees explain that the extent to which their gender identity is seen to be consistent with the imagined sex of the body – the degree to which the latter reliably signifies the former – by onlookers is the overriding factor in determining one's access to gendered toilets.

My methodological rationale is based on the premise that it is fallacious to posit an absolute difference between gender and sexuality, particularly when such a distinction is not felt to be true for everyone. While many of us may feel gender identity and sexuality to be separable and distinct features of everyday life, they are not so easily distinguishable throughout history and across cultures – our own present-day Canadian and American cultural contexts included. At the same time, I wish to do justice to the specificities of transphobia and homophobia that are, at least to some degree, separable and distinct modalities of intolerance. 'Clearly, the recognition that "gender" encompasses far more than sexual desire, and, concomitantly, that "sexuality" and sexual desire do not always align in conventional ways with gender identity, is a vital one' (Valentine 2007, 62). Trans studies, as I discuss below, is founded upon a wish to recognize the centrality of gender embodiment apart from sexuality studies, and queer theory in particular, the latter of which has been significantly less attentive than trans studies to the material conditions of trans lives and to cissexism (Serano 2007, 12).

My intention is to consider how the gendered spatial design of the public bathroom is dependent upon a cissexist and heteronormative ideal and the various ways this design impacts upon LGBTI people.

Because we all identify differently in the domains of gender and sexuality, a discussion about terminology is in order. I use 'trans' and 'transgender' interchangeably throughout the book and in the broadest possible sense to include those who defy, by choice or necessity, conventional sex and gender systems. As Stephen Whittle (2006) notes, a trans identity can

> encompass discomfort with role expectations, being queer, occasional or more frequent cross-dressing, permanent cross-dressing and cross-gender living, through to accessing major health interventions such as hormonal therapy and surgical reassignment procedures. It can take up as little of your life as five minutes a week or as much as a life-long commitment to reconfiguring the body to match the inner self. (xi)

Because some theorists and interviewees prefer to distinguish between those who are 'transgender' and those who are 'transsexual,' it should be noted that those who identify as transsexual often seek medical intervention and/or hormones to change sex, and their commitment to reconfiguring the body is very often 'lifelong.' Those who identify as 'transgender' do not always seek medical intervention. For my purposes, the designation 'trans' or 'transgender' is employed as an overarching, and necessarily imprecise way to denote those whose gender identities are, in some way, at odds with conventional sex/gender systems and to denote those who have transitioned or who are in transition.[13] I use the terms 'trans' and 'transgender' to include those who are transsexual along with those who are gender-variant, including butches;[14] those who are masculine and female;[15] effeminate men; trans queer femmes; and genderqueer, two-spirited, and transgenderist people. As Stephen Whittle puts it, the category 'trans' is 'accessible almost anywhere, to anyone who does not feel comfortable in the gender role they were attributed with at birth, or who has a gender identity at odds with the labels "man" or "woman" credited to them by formal authorities' (ibid., ix). I do, however, heed Julia Serano's (2007) warning that the trouble with the term 'transgender' is that it cultivates a one-size-fits-all approach that can obscure the particular struggles of those who face different and 'multiple forms of gender-based prejudice' (3). Throughout the manuscript I take care not to posit a universal trans subject but to illustrate the different experiences had by those who have a non-normative, cross- or transgender affiliation.

Of course, the boundaries between the various identity categories are slippery, political, and subject to contestation in academe and in LGBTI communities outside the university. I use the terms in a provisional way without intending to preclude or to guard against re-articulations and reformulations by others. The danger of using an umbrella term is that it posits a uniform collectivity and cannot do justice to the myriad differences subsumed into the category. I urge readers to consider the differences within the analytically employed transgender frame by carefully attending to the multiple voices quoted in the chapters to follow.

Those who identify as transsexual often have sex reassignment surgeries, take hormones, and/or have (or are working to change) their birth certificates, legal documents, and other identification papers to reflect their gender identities. 'Many transgendered people feel they are not the gender they were assigned and are not comfortable with their birth sex; beyond that, they feel varying degrees of identifications and belonging to another gender category' (Nataf 2006, 447). Some who transition from male to female (MTF), or from female to male (FTM),[16] identify as transsexual and not as transgender, and I respect this distinction when referring to individual interviews.[17] I do, however, include those who identify as transsexual when I reference 'trans' and 'transgender' people or communities more generally. My intention is not to erase or to minimize differences between those who are transsexual and those who are transgender (or genderqueer), but rather to be inclusive of all who have either ambivalent relations to binary gender codes or who do not identify with their gender assignment at birth and have consequently transitioned or are currently in transition.

As Susan Stryker (2006) notes, the term 'transgender' is 'more complex and variant than can be accounted for by the currently dominant binary sex/gender ideology of Eurocentric modernity' (3). In other words, there is a necessary gap between the category and the subject that cannot be made transparent in modern discourses about gender identity, sex, and sexuality. Referring specifically to the emerging field of transgender studies, Stryker also notes that the field of inquiry is

> concerned with anything that disrupts, denaturalizes, rearticulates, and makes visible the normative linkages we generally assume to exist between the biological specificity of the sexually differentiated human body, the social roles and statuses that a particular form of body is expected to

occupy, the subjectively experienced relationship between a gendered sense of self and social expectations of gender-role performance, and the cultural mechanisms that work to sustain or thwart specific configurations of gendered personhood. (ibid.)

While trans studies has distinguished itself from sexuality studies, there is a relation – however unstable, turbulent, and contested – between the two fields of inquiry. Queer theory is committed to the study of sexual norms engendered by compulsory heterosexuality. It is also focused upon how identities are disciplined by medical, legal, political, and social institutions that presume, as they also enforce, a dominant heterosexual order.

Given that gender is disciplined, in part, through the homosexual injunction and that sexuality is disciplined, in part, through essentialist (and sometimes transphobic) ideas about gender, it is not surprising that there is a close affinity between queer theory and trans studies. Because trans people can be homophobic; and gay, lesbian, and bisexual people can be transphobic, it is important to remember that the LGBTI people interviewed in this book (along with the greater heterosexual and cissexual population at large) are not always exempt from the aggressive disidentifications subject to analysis in this book. There is also, as I will demonstrate in the chapters to follow, a means by which those who are transgender are subject to homophobic assaults, and those who are recognizably queer are subject to transphobic injuries: the two mechanisms of intolerance sometimes overlap in the public milieu, as they are also, in other moments, separate projective and defensive systems leading to intolerance and hate. As Viviane K. Namaste (2000) explains:

> In Western societies, gender and sexuality get confused. For example, when a fifteen-year-old boy is assaulted and called a 'faggot,' he is so labeled because he has mannerisms that are considered 'effeminate.' He may or may not be gay, but he is called a 'queer' because he does not fulfill his expected gender role. A young girl can be a tomboy until the age of eleven or so, but she must then live as a more 'dainty,' 'feminine' person. If she does not, she may be called a 'dyke' – again, regardless of how she actually defines her sexual identity. (140)

Because trans folk are sometimes read as 'gay' (when they may, in fact, be heterosexual), it is not always possible to ascertain if it is a trans subjectivity or a perceived queer subjectivity that ignites aggression. For trans people who are gay or bisexual, the distinctions between transphobia and homophobia may be less important than their points of convergence.

I am in agreement with Valentine (2007), who argues that in the present American (and I would add Canadian) context, although the 'separation of gender and sexuality makes sense of many contemporary people's senses of self, there are also contemporary gendered/sexual subjects whose senses of self are not accounted for by this distinction' (62). In order to respect both groupings of people (and their experiences of transphobia and/or homophobia), I employ queer theory *and* trans studies throughout the manuscript. Queer theories are often focused upon desire and sexuality. Transgender studies are usually centred upon sexed embodiment and gender identity. This study requires both fields of exploration. Each field of study is grappling, albeit in different ways, with the disciplining of the body, its genitals, its gender identifications, sexuality, modalities of love, affiliations, and desires. By segregating trans and queer theory we are in danger of losing sight of the nuances, complexities, and interlocking disciplinary devices through which we are subjugated to networks of power.

Susan Stryker notes that transgender studies have a close, albeit sometimes strained, relation to queer studies (2006, 7). Because there is an intimate relationship between gender and sexuality (even as one cannot be determined or presumed by the other) heteronormativity is dependent upon conventional sex/gender systems. 'Sexual object choice, the very concept used to distinguish "hetero" from "homo" sexuality, loses coherence to the precise extent that the "sex" of the "object" is called into question, particularly in relation to the object's "gender" ' (ibid.). Of course, trans people are often very clear about their sexual orientation (gay, bisexual, or straight), but the point is that their status *as transgender* affects the way their sexualities are read by those who see bodies through a heteronormative template that presumes *non*-trans identifications, consequently erasing trans histories and trans-specific entries into heterosexuality (or homosexuality).[18]

I use the term 'queer' in an equally expansive way. When referring to 'queer people,' I have in mind all those interviewees who identify as lesbian, gay, bisexual, and, of course, as queer. The term also includes trans people who are gay, lesbian, bisexual, or queer. While not everyone identifies their sexuality as 'queer,' per se, I use the term as a means to denote those who are not heterosexual. When quoting interviewees,

I use their chosen sexual identifications where relevant unless they are not given on the demographic questionnaire. I am also careful to distinguish between homophobic and transphobic bathroom encounters when such distinctions are evident. I do my best to articulate the differences between what queer and/or trans people say about bathroom encounters. The study is premised upon the assumption that identifications do not necessarily delimit or determine what one does or does not observe or experience in the bathroom.

There is, as with any interview-based project, a politics as well as a methodology (stated or unstated) that guides my interpretations of interview data. My data are based on interviews conducted with LGBTI people. Open-ended questions were asked about a variety of experiences in bathroom spaces having to do with gender panic and sexuality. Because qualitative research on gender, sexuality, and bathroom accessibility is relatively scarce, I intended to capture a wide range of perspectives on the polemics of gender recognition in toilets. I employed six research assistants to aid in the interview process. Together, the seven of us interviewed 100 people. While we each asked standard questions, each interviewer modified the questionnaires to fit the unique experiences and testimonies offered by individual participants. Most interviews took place in large Canadian and American cities. Because of funding constraints, the vast majority of interviews were conducted in Toronto, where I live and work and where I recruited my research assistants. It should also be noted that there is a disproportionate focus on facilities in large urban cities as opposed to small rural towns and communities. There are, however, occasional references to restrooms in rural areas when interviewees who reside in large cities recount experiences of travelling through less densely populated areas or relay memories of having lived in small-town communities in Canada and the United States.

Bathroom encounters discussed in the interview data are most often in gas stations, bus and transit stations, airports, bars and restaurants, malls, workplaces, schools and universities, parks, and community centres. The research focuses on public or semi-public bathrooms and not private, in-home bathrooms, which are usually gender neutral. The primary goal of this research is to understand how gender-segregated toilets in Canadian and American public spaces discipline gender. Because public toilets in these countries are relatively generic (family bathrooms along with those designated for persons with disabilities notwithstanding), I did not focus on differences between particular toilets – their management, design, location, ownership, and the people who typically use them. While I regard these differences to be significant and (if known) flag them, where relevant, to the interview data, I intend to focus on what is common to North American bathrooms – namely, their gender-segregated designs. My primary interest is in the relatively generic architectural designs and the choreographies they engender. See, for example, Kira (1966, 1976), who outlines the body choreographies structured by lavatory designs that are of central concern to this study. While his work on New York City toilets pre-dates this study, the designs and eliminatory choreographies he subjects to detailed analysis have not changed in any significant way and are, in fact, common throughout North America.

I use the words 'bathroom,' 'toilet,' 'washroom,' 'lavatory,' 'restroom,' and 'facility/ies' interchangeably throughout the manuscript because they are all words used by interviewees. I caution readers to remember that these terms are all imprecise referents and that there is, as noted by Kira (1966, 1976) and a host of others who study the language and cultural politics of elimination, no one word (that is not itself a euphemism) to denote the room under investigation. I also use 'sex-segregated' and 'gender-segregated' to mean the same thing. I place quotation marks around the words 'men's' and 'women's' preceding references to toilets, in order to flag the gendering of the restrooms and to de-essentialize the gendered spatial composition of the toilet.

Interviews were on average one hour and fifteen minutes in length. Participants offered novel, provocative, and in many other ways unconventional testimonies. We interviewed people with visas and temporary work permits; those who are under-housed and underemployed; those who had migrated to Canada and to the United States; those who had not received higher education; and those who struggle with addictions and mental health. We interviewed sex workers, custodians, people who are physically disabled, those who live with a mental-health diagnosis, those who have AIDs or are HIV positive, diabetics, artists, activists, professors, students, writers, and community organizers. While questions were asked about housing and employment status, it was difficult to obtain reliable indicators of class and income because of the large number of interviewees who were students and thus underemployed and peripatetic in pursuit of affordable housing. I give participant demographic information the first time each subject is introduced to readers. I also provide reminders of gender and sexual identities, along with other particulars, as they are relevant to quotations in later chapters when a given interviewee is reintroduced.

The data are limited to the extent that the vast majority of interviewees are white, able-bodied,[19] middle- to upper-class graduate students who are often politically active and committed to LGBTI rights. A concerted effort was made to interview people of colour, people of diverse ages, and relatively non-politicized LGBTI folks; however, those who responded to our calls for participation tended to be white, in their late twenties or their thirties, and well integrated into queer and/or trans communities. There is, at the same time, significant diversity among the interviewees. The youngest participant is eighteen and the oldest fifty-nine. Interviewees included people with cultural roots in East Asia (China, Hong Kong, and Japan); the Middle East (Lebanon, Iran, Iraq, and Israel/Palestine); South Asia (India and Pakistan); Ireland; the United Kingdom; western Europe (France, Italy, Germany, Portugal); northern Europe (Finland, Norway, Sweden); as well as First Nations people (Ojibway, Cree, and Navajo) in Canada and the United States. All of them resided in and/or had travelled to and/or worked in Canada and/or the United States at the time of the interview.

The study includes a comparable number of participants who identify as transsexual or as transgender; as genderqueer; and as gay, lesbian, or bisexual. There are at least three self-identified intersex interviewees. A few participants identified as transgenderist. There were, however, slightly more trans men and transmasculine people than trans women and transfeminine people in the study. It should also be noted that most transsexual participants were either in transition or newly transitioned. If the study had involved a greater proportion of post-transition transsexuals, there might have been fewer testimonials about the exclusionary designs of toilets. Equally, there might have been fewer reports of gender-based interrogation if it is true that those post-transition are more skilled in navigating gendered space and more clearly recognizable as the gender to which they identify. It is likely that transsexuals interested in participating in the *Queering Bathrooms* study were also those who were at the time contending with bathroom-access issues because they were newly transitioned or in transition. It should also be said that many trans men interviewed in this study use the 'women's' room, as opposed to the 'men's' room. While many trans guys in the general population regularly use the 'men's' toilet (and often without difficulty or trepidation), my sample may include a disproportionate number of trans men who do not. This is, very likely, because most trans-guys interviewed were in transition or newly transitioned and therefore cautious and concerned about safety and accessibility in the 'men's' room.

It has been suggested by astute proofreaders of earlier drafts that some of the interviews may reflect fears about the dangers of the 'men's' toilet that are overstated or imaginary. This is particularly true for those who speak about but have not actually used the 'men's' room. I therefore want to gently caution readers against reading the interview quotations as transcendental truth-claims. In many cases, what people say about the gendered composition of the toilet is shaped by dominant discourses and imaginings about gendered toilets, rather than being unmitigated and transparent accounts of the so-called real. In some cases, interviewees give voice to cissexist and homonormative ideas about what men and women do and don't do in toilets that reveal levels of intolerance and prejudice comparable to those among people who are not LGBTI. While my use of the interview data may sometimes appear to suggest that it is only cissexuals and heterosexual people who are intolerant of LGBTI folk, it should be noted that interviewees themselves can assume homophobic and transphobic positions. It must be stressed that not all trans people have difficulty using gender-segregated bathrooms. It would be false to presume an absolute difference between those who are LGBTI and those who are not.

Interviewees were selected using informal networks, personal contacts, and referrals from colleagues and community-based, activist, mainly university groups committed to trans and/or queer-positive politics. A variety of strategies and techniques were used to recruit people to the study, including postings, community advertisements in the queer press, list-serves, formal announcements, and individual invitations. The interviews typically took place in coffee shops, restaurants, libraries, classrooms, university offices, and in private homes when appropriate. All interviews were taped and transcribed. I use pseudonyms for participants unless they requested that I use their real names or chose to remain nameless.[20]

Because it is impossible to do justice to all the insightful and provocative testimony included in the interview data, this book is limited to an excavation of the interviews (and sound bites) relevant to the government of gender and sexuality. My interest in queer theory, trans studies, and psychoanalysis informs the way I read and interpret the interviews: what I highlight and place in the foreground, what I quote or ignore, what I theorize or forget to mention. My analysis is also informed by the literature, by countless sleepless nights reading the interview transcripts, and also by my own experiences in gendered toilets. I am not transgender. I am a white, able-bodied, queer femme professor (with tenure), living in Toronto, and have always been interested in what feels to me to be a strange and antiquated division of gender by urinary design. I access the 'women's' room easily and without question or interrogation by onlookers. My lovers and partners (most of whom identify as butch) do, however, experience such inquisition, and I have seen – from the vantage point of an ally – the incredulous and prying stares often given to trans people in the 'ladies' room.

It should be clear by now that the interpretations offered in this book are my own. Although I hope to give voice, recognition, and affirming space to those who have been interviewed, the book does not document a knowable, transparent, and unequivocal truth – as if any book ever could. There are many fascinating stories and narratives that do not appear in the final version but that informed my readings of what has been included and analysed in this book. Strangely – or, rather, queerly – I was especially touched by many stories that did not make the final version. As with any qualitative project involving large population samples, extensive interview data, and opportunities for intersectional analysis, it is simply not possible to include everything. I did, however, do my best to include something of each interview in this book. Some interviews were highly analytic (and easy to quote in an academic text), while others were especially poignant and moving *because* they gave voice to the mundane, banal, everyday reality of encounters in toilets. Other interviews were memorable because of the effects transmitted through intonation and discerned by the interviewers through intuition and sensorial registers (but not easily quoted). Still others were marked by trauma, trepidation, anxiety, and fear of what would be made of and done to their words in the final analysis. Some interviewees, with keen political insight, were able to clarify the mechanisms through which LGBTI people are disciplined and subjected to gender and/or sexual panic. In choosing bits and pieces – quotable sound bites, so to speak – I tell not one but multiple stories. A collage of adventures and injuries, insights and traumas, desirous and awkward encounters, this book is an attempt to bring together a myriad of LGBTI voices on the subject of the toilet.

Outline of Chapters

In chapter 1, I argue that the history and cultural politics of excretion are relevant to gender and sexuality studies. The late-eighteenth-century worry about disease and hygiene shaped the bio-political interest in gender and sexual propriety. This is evident in the metonymic association between clean toilets and sexual morality, dirty toilets and sexual immorality, along with an unprecedented gendering and privatization of elimination in city space. I also introduce the concept of the bodily ego to make an argument about how gender is employed to seal the body and to offset worries about disease and human mortality (often projected onto toilets).

Chapter 2 investigates the problem of gender misreadings in bathrooms as evidenced in the interview data. Using Butler's (1997, 2004) analytic of gender and Kosofsky's (2003) notes on being effaced in public mirror circuits, along with trans studies on what it means to be rendered invisible (Namaste 2000), to lack cissexual privilege (Serano 2007), and to feel trans-specific forms of rage (Stryker 2006) when one cannot abide by the 'norms of gender embodiment' (253) authorized in public space, I outline the injuries to self incurred by the gendered spatial designs of the public toilet.

In chapter 3, I use Michel Foucault's (1979) theory of panoptic power – involving the use of light, optics, and surveillance – to consider how toilets are designed to visually apprehend gender. Interviewees note how a gap between gender identity and the sex of the body, as intercepted by others, is subject to microscopic attention. Focusing on the use of mirrors, fluorescent lights, reflective surfaces, white walls, and panoptic eyes (nowhere and everywhere), I consider how visual economies of power operate to discipline gender.

In chapter 4, I use Kaja Silverman's (1988) analytic of the acoustic mirror to theorize how a lack of synchronization between body sounds and images puts gender at risk. Interviewees confirm that the architecture and design of the lavatory deploy visual *and* acoustic mirrors to govern gender. Panopticism works in harmony with acoustic registers. Cissexual laws of symmetry require masculine and feminine subjects to assume divergent urinary positions. How one stands or sits, hovers or squats, indicates gender. The urinary echo orchestrates a truth about the body and its genital composition.

Chapter 5 adds to the literature on social abjection (Kristeva 1982, Lahiji and Friedman 1997, McClintock 1995, Thomas 2008) by considering how LGBTI people are culturally coded as abject. Interviewees testify to how their bodies are sometimes read as polluting, in need of symbolic clean-up and removal in the gender-normative landscape. White, sterile, industrial bathrooms in North America are rooted in colonial and puritanical angst about racial and class mixing dating back to the eighteenth century. Worries about social difference are now – in the present-day Canadian and American washroom – about gender purity, its legibility in the public domain, and this purity is mediated by dis/ability, 'race,' and class.

In chapter 6, I show how heterosexual matrices are mapped onto toilets. Gender panic is often fuelled by worries about sex acts that are hedonistic, queer, undomesticated, and unproductive. As non-trans gay male interviewees note, homophobia is often focused upon the icon of the sodomite. Licentious sex in bathrooms is subject to prohibition because it confuses gender and is thought to be unresponsive to domestication by marriage. Gender and sexual regulation operate in concert. By forging erotic bonds in public bathrooms, interviewees build what Berlant and Warner (2002) call 'counter-publics' (199) in defiance of heteronormative house rules.

I conclude that the toilet disciplines gender by imposing a rigidly gendered and heteronormative plumb line. Interviewees give voice to how we may redesign the room so that it does not mirror and consolidate a normative gender order. By imagining queer and trans-positive designs, ones that can accommodate a range of bodies, desires, genders, and non-normative identifications, we may build a more inclusive and luxurious bathroom. The project of the twenty-first century is to build a fluid and less exacting plumb line, one that does not mirror an essential connection between sex and gender, body and identity, but makes room for new and unexpected configurations of gender and sexed embodiment in space.

1 Queering Bathrooms: Gender, Sexuality, and Excretion

> [The] subject of toilet styles and toilet behaviour has received far too little attention as a research topic, and observations of social practices are few ... Yet defecation and self-cleaning, like procreation and food consumption, are an inextricable part of the human condition, regardless of race, ethnicity, class, age, gender, political beliefs, religious beliefs, or sexual orientation and should, in my opinion, invite more serious study.
>
> (Srinivas 2002, 369)

Michel Foucault (1978) tells us that the history of sexuality is about how sex became a secret to be exploited, a truth or window into the soul to be discovered and diagnosed. Sexologists from Sigmund Freud to Alfred Kinsey complained that people ridicule sexuality studies because they were (and continue to be) regarded as in bad taste, unseemly, depraved, and slightly perverse. What we do (or don't do) in bed was not a matter of legitimate academic inquiry. Even today sexuality studies remain marginal to the more legitimate disciplines in the social sciences and humanities. Nor is sexuality an appropriate subject of public discourse. There is something crude and mundane about the body and its sexuality. The sexual instinct is imagined to be over-determined by nature and the supposed biological difference between men and women. There is nothing to say about sexuality precisely because it is thought to be a biological given – we have it, we do it, and it is essential to the reproduction of the species. Academic inquiries about the body, its desires, and its sexual practices are often dismissed as vulgar or irrelevant.

The same can be said about the history and cultural politics of excretion from the Victorian era to the present-day North American context.

The management of elimination and the body's fluids and orifices in the modern lavatory are part and parcel of the bio-political regulation of the body, its gender, and its sexuality from (at least) the Victorian era through to the present day. The institution of the gender-segregated toilet, now commonplace in North America, dates back to eighteenth-century London and Paris. It is a little-known fact that the first gender-segregated public toilets in Europe were assembled in a Parisian restaurant for a ball held in 1739. The organizers of the ball allotted '*cabinets* with *Garderrobes pour les hommes*, with chambermaids in the former and valets in the latter' (Wright 1960, 103) to ensure a proper division by gender. The segregation, first implemented by the Parisian upper classes,[1] was intended to accentuate sexual difference and to project its difference onto public space. Gender-segregated lavatory design in public was, in its original incarnation, meant to indicate class standing and genteel respectability.

Differences between genders were increasingly exaggerated by British and European public toilet facilities in ways that were class and culturally specific and driven by a heteronormative imperative. Portable female urinettes made in the early 1700s, in London, were of glass, leather, or ceramic (to be carried around on one's person), while male urinals were permanent concrete structures built in public city space. Eventually, public toilet facilities were built for women in the late 1800s and early 1900s. Although public lavatories for women enabled the so-called 'genteel' sex to frequent the city streets of London and Paris for longer periods of time (and to shop, as retailer Timothy Eaton of Toronto insisted women needed to do for capitalist accumulation and profit in the early 1900s), they functioned to create a rigid, architecturally imposed gender divide that is still with us today. There is an under-studied parallel between thought about gender and sexuality and the privatization of excretion in the Victorian era exemplified in the development of the now common flush toilet. The ordinances governing the management of excretion in London, along with the technologies of the water closet developed by a host of sanitary engineers, plumbers, and inventors of the eighteenth century, led to a historically unparalleled privatization and gendering of the eliminatory function. The present-day problem of restroom segregation by gender is attributable in no small measure to the curious ways in which Londoners and Parisians imagined and gave rise to toiletry provisions in city spaces. The 'lavatory as we know it today was invented some 100 years ago. Since the 1880s it has changed neither its workings nor its basic shape' (Lambton 2007, 25).

While sexuality was subject to repression in the Victorian age (according to Freud), *and* transformed into discourse, deployed in fields of disciplinary power, and used to govern populations (according to Foucault), it seems that elimination is 'beyond the pale' and subject to visceral disgust in ways that sexualities never were. This may be because 'shit, turds, crots, ordure, deposit, fecal matter, excrement, droppings, fumets, motion, dung, stronts, scybale, or spyrathe' (Anspaugh 1994, 4) disable the imagined boundaries between inside and outside, pure and hygienic, impure and disease-laden, self and other. Body boundaries are felt to be unstable in the face of shit, and identificatory injuries are incurred because it symbolizes 'dirt,' 'disgust,' and 'defilement,' body 'matter out of place' (Douglas 1966). If the repressive hypothesis is a red herring, then the impulse to quell talk and study about modern lavatories is, similarly, misguided and less than innocent.

Great pains are taken to secure a normative cissexual (non-trans) gender order and to police and map the coordinates of that order in bathrooms and bedrooms. Stifled conversations, hushed tones and whispers, body squeamishness, and looks askance all function to attend to what goes on in bathrooms without seeming to. This is a generative quiet; one that conceals as it orchestrates a truth about the body, its gender, and its genitals. In other words, the toilet has a taboo and 'unofficial' story to tell. Bathroom history – from the Victorian period to the twenty-first-century Canadian and American context – has much to say about the social coordinates through which bodies are disciplined by normalizing regimes. The bathroom is a repository for the societal unconscious: all that Western modernity forgets, disavows, and casts asunder in polite discourse. As Victor Hugo wrote in *Les Misérables*, 'A sewer is a cynic. It tells all.' The Parisian sewer, for example, gained a reputation for subterfuge: 'Crime, intelligence, social protest, freedom of conscience thought, theft, all that human laws prosecuted or have prosecuted, was hidden in this pit ...' (Hugo quoted in Horan 1996, 92).[2] The sewer became a repository, or rather a barometer, of the societal unconscious right up to the French Revolution, after which the Parisian sewer system was subject to clean up and presented as an example of modern utility, sanitation, and efficiency.

In Freudian psychoanalytic theory, the unconscious is a formation and a system through which the individual banishes antisocial thoughts and anti-Oedipal desires from consciousness. 'In Freud's work, the unconscious is another scene, a parallel process which works by its own logic; it uses its own language, signs and symbols, makes its own connections; it is born out of prohibitions, repressions and taboos – all of which are nested in the psycho-social-spatial field of everyday life. The unconscious is a zone of primal exclusion, it is a wound, but it knows no Law, no negation, everything is permissible – it is mythic and magical – yet it is as real as anything else' (Pile 1996, 76). In other words, the unconscious, like the toilet, is a dumping ground for unacceptable impulses, practices, identifications, and desires. 'Plumbing, so often aligned with bodily trauma, is a volatile signifier of that which cannot directly be acknowledged in the symbolic order – a toilet, a plunger, a shower stall to take the place of the unspeakable – and to make it all the more charged' (Morgan 2002, 178).

How the orifices[3] of the body – its fluids and sphincters, its urinary positions (standing or sitting), and its conformity to gender signs on bathroom doors – are disciplined speaks volumes about gender and the bio-politics of government. It also tells us much about what has been forgotten in the making of the modern gendered body. There seems to be a fetishistic quality to the obsessive interest in the gender of bathroom users. Separating bodies by urinary capacities – real and imagined – is a way to ensure sexual difference when our bodies do not always lend themselves to absolute and exacting divisions by gender. If shit is the great equalizer,[4] and women can stand to pee, and men are, on more than a few occasions, known to sit, then elimination (not unlike homosex) may confuse sexual difference.

Apart from the Victorian bedroom we would be hard pressed to find a room more saturated by taboo and prohibition than the modern lavatory now institutionalized in American and Canadian cities. Both architectures – the marital bed and the lavatory – are designed with sex and genitals in mind. They shape how bodies may come together and how they are kept apart. Each room is preoccupied by the spectacle of sex. A heterosexually specific and reproductive morality is employed to set parameters on how, when, where, and in what manner body fluids are evacuated, by whom, and into which orifice (in the case of sex) or receptacle – urinal or porcelain oval (in the case of the toilet). Each room comes equipped with gender-specific signs and rules of entry. Men and women come together in a marital bed and are kept apart in a public lavatory. What happens in each room is a public curiosity – a secret engendering ideas about bodies, sexual practices, genitals, and clandestine desires.

Public censors in the modern era are productive. By gossip, innuendo, fantasy, and deduction we imagine a truth about gender and its sexual practices in bedrooms. In bathrooms we imagine a concordant truth, one that is, often, more licentious and perverse. We are either innocent – unmoved by, or unresponsive to, the fluids evacuated by others – *or* queerly animated by orifices and fluids shed in 'dirty' places. There is no shortage of sexual practices catalogued by sexologists having to do with elimination and anal eroticism. One only need refer to *Psychopathia Sexualis: A Medical-Forensic Study* (1965) by Richard von Krafft-Ebing, to Sigmund Freud's (1975 [1905]) writing on anal eroticism and psychosexual development, to John G. Bourke's (1891) *Scatalogic Rites of All Nations*, or to Brenda Love's (1992) *Encyclopedia of Unusual Sex Practices* to be persuaded that the history and cultural politics of elimination are related to human sexuality. Perhaps the most obvious connection among the management of excreta, gender, and sexuality is to be seen in the design of early modern urinals for women. The London Science Museum houses early-eighteenth-century female urinals made of glass in the shape of an erect penis and testicles (Penner 2005). In an interesting study of the glass urinals, Barbara Penner (2005) notes that the 'spectre of sex hangs over my enquiry' (87). At the time of writing, there is even a female urinal on display at the Museum of Sex in New York City (designed by Alex Schweder, 2001). In her pictorial excavation of Britain's lavatories, Lucinda Lambton (2007) writes that she was surprised to learn that the 'Gentlemen's Cloakroom of the Manchester Club' (designed in 1870 and equipped with washbasins) had been beautifully restored by Agent Provocateur, which turned it into a rather risqué lingerie store (with naked female mannequins, BDSM gear, and fashionable corsets).

While we may project a lewd and licentious past onto the boudoirs and lavatories of the Victorian era, there still remains an unexamined *Victorian* preoccupation with the body – its gender, its genitals, and its excreta – that is part and parcel of the history of the water closet bearing on the present-day North American context. Both the bedroom and the bathroom are over-determined by heterosexual angst (and excitation) about illicit border crossings. Referring to the early-twentieth-century American context, Patricia Cooper (1999) writes that

> concerns about bodies, sexual propriety, decorum, and morality arise in bathroom discourse ... The very creation of bathroom spaces, which are routinely separated by sex, reflect cultural beliefs about privacy and sexuality ... The bathroom is a place where genitals may be touched but not primarily for the purpose of sexual stimulation ... So women's and men's

bathrooms assume heterosexuality and the existence of only two sexes, permit genital touching, and reject overt sexual expression. (25–6)

Who may enter the private geographies of the bedroom and semi-public bathroom is determined by ideas about gender and genitals.

Gender is mapped onto public space through public toiletry provisions. French psychoanalyst Jacques Lacan coined the phrase 'urinary segregation' (2006) to denote how the masculine and the feminine are, primarily, linguistic signifiers, as opposed to biological entities. The signs 'Ladies' and 'Gentleman' on toilet doors are instrumental to the dominant gender order in the West. As a gendered architecture of exclusion, the washroom is beholden to the discourses of masculinity and femininity – as oppositional and mutually exclusive locales – anchoring so-called binary and unchanging subject positions.[5] In her discussion of gender and architectural design, Catherine Ingraham (1992) asks of the now infamous signs 'Ladies' and 'Gentlemen' theorized by Lacan, 'In what sense ... are the doors immunized (in their formal sameness) against the difference that the labeling argues for?' (263). Noting that modern architecture tends to sterilize and to desexualize space, Ingraham concludes that the very absence of sexuality signals its presence. Ingraham explains that by 'casting space as neutral, architecture is able to avoid the *specificity* of difference that is the very structure of sexuality, insofar as sexuality is paradigmatically about the specificity of, identity through, and competition between gender differences' (1992, 262).

Modern architectural design reflects a desire to eradicate gender difference and to sanitize sexuality. 'In general, architectural culture has kept the sexuality of space repressed, kept space sterilized as a technical economy under the control of the mythological design architect' (Ingraham 1992, 264). Gender is sanitized by the likeness of the oppositional doors. In the *History of Shit*, Dominique Laporte (2000) makes a related argument about how the hygienic treatment of body waste is at the heart of modernity. Like the history of sexuality, as told by Michel Foucault (1978), the story of elimination is not shadowed by a repressive and unproductive silence. It was, instead, deployed by sanitary engineers, city planners, architects, inventors, and hygienists to clean up (or to profit from) the social body (now coded as diseased and dirty).[6]

> Blood, milk, shit, sex, corpses, sperm, sewers, hospitals, factories, urinals – for three quarters of a century, the hygienist has spoken of these ceaselessly. He is the prince consort of bourgeois civilization, of colonialist

Europe as embodied by Queen Victoria. Excremental issues are at the heart of his accounts, memoirs, observations, reports, letters, essays, bulletins, etc. (Laporte 2000, 119)

One need only reflect upon the history of racially segregated bathrooms in the United States, upon the privatization of washroom facilities in the present day (Bess 1997; Cohen et al. 2005), and on the imposition of the English water closet on the non-Western world from the early days of colonial expansion to present-day tourism (Moore, 2008) to understand the role of excretion and its management in histories of colonization,[7] capitalism,[8] gender, class,[9] 'race,'[10] sexuality, disability, imperialism, and global inequalities.

Despite the centrality of the modern water closet to western-European modernity, it is curious that there have been relatively few scholarly investigations into the cultural, political, and economic significance of the toilet. Social scientists frequently mention the dearth of research on the subject, but sustained scholarly attention to the lavatory as a topic of investigation in its own right is scarce at best. But let us begin with the exceptions. In the mid-twentieth century, social thinkers referred to the lavatory as a place where both taboo and moral values pertinent to the body could be seen. For example, in 1955, American anthropologist Horace Minor commented in jest that 'many of the rituals that behaviorally express and sustain the central values of our culture occur in bathrooms' (Cahill et al. 1985, 33). In *The Presentation of Self in Everyday Life* (1959), Irving Goffman writes that defecation upsets public performances and rituals through which we present ourselves as clean and pure. Spencer Cahill observes that while bathrooms reveal much about the theatre of society, 'students of everyday social life have shown little interest in this topic of inquiry' (Cahill et al. 1985, 34). As Annabel Cooper and colleagues (2000) summarize: 'Behind the apparently mundane nature of these "small" spaces lies a history of the sexed and gendered body in urban space, the civilizing and modernizing of landscapes and bodies, the nature of citizenship, and the shifting practices and technologies of private and public' (417).

Cavalryman John G. Bourke's exhaustive study of excremental practices the world over, published in *Scatologic Rites of All Nations* (1891), leads even the most sceptical reader to acknowledge a pivotal relationship between coprophagic rituals and what was, in Europe and the Americas, called civilization. Bourke catalogued a broad assortment of anthropological and ethnographic data to support the conclusion that excretion and defecation play central roles in world religions – their belief systems, symbolisms, and practices – from the Middle Ages to the dawn of the modern era – not to mention the place of urine and scat in food preparation, medicine, courtship, sex play, reproduction, theatre, dreams, art, festivals, mythology, the building of bridges, housing, modern urban building constructions of all sorts, farming, fishing, agriculture, industry, mourning and mortuary ceremonies, and, finally, the 'Feast of Fools' of ancient Europe.

Margaret Morgan (2002) claims that the toilet is the 'icon of the twentieth century' (171), and yet it is not well theorized in social and political histories of modernity. The toilet is, she writes, a centrepiece, a vestibule of modernity and its investments in cleanliness, utility, technology, hygiene, industry, and cleanliness.[11] Noting that modernity is, in part, marked by the fall of religion and the rise of science – or rather the use of science *as a religion to* manage bodies and cultures in the service of cleanliness, health and hygiene, and morality – Morgan contends that the Victorian water closet epitomizes the transition from pre-modern to modern. Today, we worship the modern toilet, its cleansing and sanitizing technologies, much as pre-modern Europeans worshipped religious deities and pagans. 'But if the toilet is iconic, then the sacrament is reversed: in the Eucharist we imbibe the blood and body of the Christ figure. Here, in inversion, we present our blood and shit and piss before the shrines to hygiene and modernist aesthetics. That is, if god is said to enter our bodies in the pre-modern ritual, then it is we, as gods, who enter the body of the State and Metropolis in the modern ritual that is the adoration of the cubicle' (Morgan 2002, 171). Referring to comparisons made between urinals and Madonnas or Buddhas, to functional and durable sewers as 'noble' and 'holy,' and, also, to John Wesley, who said that 'cleanliness is next to godliness' (Quoted in Morgan 2002, 171), Morgan notes how hygienic practices (so many of which take place in bathrooms – private and public) became modern religious practices bound up with identity maintenance.[12]

Laura Kipnis (1996) observes that over the course of the modern period previously 'communal activities – sleeping, sex, elimination, eating – became subject to new sets of rules of conduct and privatization. An increasingly heightened sense of disgust at the bodies and bodily functions of others emerged, and simultaneous with this process of privatization came a corresponding sense of shame about one's own body and its functions' (135). Good manners came to be about bodily integrity and containment. Sex and elimination were subject to confinement (the bedroom and the water closet). Kipnis argues that

As far more attention came to be paid to proprieties around elimination, to hygiene, to bodily odors, and to not offending others, thresholds of sensitivity and refinement in the individual psyche, with the most shameful and prohibited behaviors and impulses (those around sex and elimination) propelled into the realm of the unconscious. (Ibid., 136)

The proximity of the sexual, urinary, and excremental orifices in women led many to quadrant and to specify the uses and functions of otherwise interchangeable erotic zones and orifices. In the otherwise anatomically correct drawings of the penis composed by Leonardo da Vinci there are two separate passages mapped out, one for semen and the second for urine (Friedman 2001). Certainly the 'migration of the tub from public facilities to various rooms within the house ... [and eventually to] private and specialized room[s] in the master-bedroom]' (Braham 1997, 217) is suggestive of the attempts to privatize the body, and to separate its sexual and its urinary and defecation rituals. In many (if not all) Western cities, the 'nineteenth century saw public urination become indecent, and "an altered threshold of public decency eventually cordoned off the public toilet as the only appropriate place for such bodily function" ' (Cooper, Law, and Malthus 2000, 418–19). As a result of the perceived similarity (or slippage) between sex and elimination, the toilet (like the bedroom) figures prominently in the history and present-day cultural politics of gender and sexuality.

Toilet training is often remarked upon in psychoanalytic writing, and in Freud's theories of psycho-sexual development in particular. In *Civilization and Its Discontents*, Freud notes that the English water closet was a symbol of beauty, cleanliness, and order: modern fixations and measures of civilization. In his discussion of anal eroticism, he (1960 [1917]) wrote that the 'concepts of *faeces* (money,[13] gift), *baby* and *penis* are ill-distinguished from one another and are easily interchangeable' (296) in early psycho-sexual development.[14] He further elaborates upon what he sees to be a relationship between toilet training and gender acquisition. The Oedipal complex, which, according to Freud, begins with the onset of toilet training, straightens the body out, its appropriate genital configuration, objects of desire, and taboo zones. Toddlers know little about which orifices are to be subject to sensation and which are to be ignored. The modern-day anus is to be extricated from the domain of the sexual. Anal cavities and faeces confuse sexual difference. Toilet training is largely responsible for the sorting out of the body's substratum: the orifices used for elimination are to be distinguished from those used for hetero-sex. The vagina not the anus, the penis not the faecal stick (coded as baby, money, or gift in the Freudian unconscious), are to be dominant centres of pleasure. Freud hypothesizes that 'The faecal mass ... represents as it were the first penis, and the stimulated mucous membrane of the rectum represents that of the vagina ... during the pregenital phase ... penis and vagina were represented by the faecal stick and the rectum' (ibid., 300). The boy learns that the faecal stick (coded as penis) is detachable, and the association instils a fear of castration. It is on the pot that the toddler learns to renounce anal pleasure (narcissistic love) for object love by doing as his or her mother wishes on the potty, and the mother is, according to the Freudian schemata, later replaced by a socially acceptable female substitute. Functional love is, in the Freudian analytic, essential not only for gender-identity development but also for participation in modern-day civilization.

Of course, one need not accept the chains of association made by Freud to recognize the relevance of the toilet and the body's mechanisms of elimination to ideas about human sexuality in the Victorian era. There are multiple references to anal eroticism, urine, and faeces in European discourses at the turn of the twentieth century. For example, in a fascinating postcolonial history of chocolate, Alison Moore (2005) notes that 'coprophagia (the eating of excrement) was most famously eroticized by the Marquis de Sade, and has been documented as a practiced sexual variation by sexologists and psychiatrists consistently from the late nineteenth century to the present' (52). The fetishization of chocolate in western-European countries was based in part, as Moore suggests, on a thinly veiled (however repressed) coprophilic tendency specific to the West. 'James Hamilton [a Jungian analyst] notes the recurrence of excrement motifs in dreams and myths and relates these to the psychic underworld, the realm in which reality is turned upside-down and inside-out, hence excrement is eaten, entering the body instead of exiting it' (ibid., 62–3).[15]

In her discussion of the history and symbolism of excretion in the nineteenth century, Alison Moore (2008) further notes that it is 'very easy for us to imagine excretion as something obscene and shameful along a similar trajectory to sex, or indeed because the notion of excretion as [sic] something even more private and embarrassing than sex ... in current-day industrialized cultures.' It should be remembered that Queen Victoria was not only unable to imagine lesbianism but was, according to Lewin (1999, 64), puzzled by toilet paper floating in the Thames. Both were unimaginable and/or indecent. In fact, the Queen did not know what toilet paper was until she asked Dr Whewell, master of Trinity

College, Cambridge, while strolling along the river bank: 'What are those piece of paper floating down the river?' The master tactfully responded: 'Those, ma'am, are notices that bathing is forbidden' (Horan 1996, 95). Lesbian sex and toilet paper were seemingly off the royal radar.

Talk about latrines was not only about gender, it was about sex. Let us not forget that the harlot (like stray toilet papers) was a worry to sanitary engineers in Paris. There were in France parallels evidenced between Parisian ordinances governing public sewage disposal and the vilification of sex work. By regulating prostitution, sanitary engineers like Alexandre Parent-Duchatelet (who conducted extensive sociological studies of the trade), believed they were 'cleaning up' the social body and keeping it free of disease and contaminating fluids: sewage and vaginal fluids (particularly those of harlots) were conflated into one overarching threat to health and hygiene (Morgan 2002, 173). 'In the 1830s the French town planner Jean-Baptiste Alexandre Parent-Duchatelet had explicitly related prostitutes to excrement' (Moore 2008, 7) and recommended the regulation and control of both. The Madonna and whore figure prominently in discourses of purity and filth.

In her discussion of noise, filth, and stench, Emily Cockayne (2007) notes a connection between sexual deviance, open sewers, and rot. She writes that 'matter that polluted the urban environment was also compared to the morally filthy. Sex and soot entwined in smuttiness' (237). When city-wide sewer systems were built in London and Paris they were compared to the sex organs. In *The London Jilt* (1683), female prostitutes were described as 'filthy, nasty, and stinking Carcasses ... A Whore is but a close-stool to Man, or a Common shoar that receives all manner of Filth' (quoted in ibid., 236–8). There were, in London and Paris, ever-present associations made between female prostitution, sexual vice, and overflowing, disease-ridden city cesspools. Female virtue was, by contrast, associated with hygienic septic systems and white china-like porcelain receptacles such as bourdalous (urinary pots made for women), decorative commodes, and close stools covered in velvet and inlaid with pearl and ribbon discreetly placed indoors. Modern advances in toilet technology were often touted in terms of hygiene and white, virginal purity.

The discourses of Parisian sanitary reform were framed by the language of sexual prohibition, whiteness, and purity. The prostitute was also identified with London's city drains and urban filth (Pike 2005, 70). She was to be symbolically flushed down the toilet after sex. Removed from sight, the prostitute was aligned with the unsightly underground.

By sanitizing the city, Londoners and Parisians also sought to reform it. Vice (sex) and dirt (excreta) were the objects of purification. 'Fallen women' were aligned with the horizontal and downtrodden. Female prostitution became a problem of engineering (Pike 2005); a relatively simple matter of segregating urban space into the visible (respectable) and invisible (disrespectful). 'Both the French and the English discourses relied on a vertical segmentation of space in which the prostitute constituted a primary threshold between above and below' (ibid., 72). City sewers were to become like the white 'ladies' of polite society, clean and decent. The underground was imagined to be a solution to sexual depravity, but, as with all 'problems' and 'solutions,' they are metonymically related and, thus, conceptually confused.

The French bidet (a genital-cleaning technology), for example, acquired a reputation for sexual immorality. While designed to clean the genitals after elimination, the bidet came to be associated with female prostitution. 'To the English and many other Westerners, the French bidet was considered a product of French immorality and used for cleansing genitals after sex' (Horan 1996, 72).[16] In his historical discussion of the water closet, Lawrence Wright (1960) similarly notes that for the British the bidet denoted sexual impropriety.

Toiletry technologies, their designs and geographical locales, tell us much about gender and who is to occupy public city space. For example, it was widely believed that respectable women should not be seen unaccompanied on Victorian streets. Ladies were to be confined to the home or escorted by a gentleman in public. Female evacuation in the Victorian era was a privatized (and sometimes disguised) event. The inner linings of Victorian dresses worn by women were often stained by urine (and such garments are now displayed in English museums) indicating that 'ladies' did use these garments to cloak the practice of urinating while standing outside in public. Before indoor plumbing, female urination tended to be discreet. 'The typical Victorian woman, it seems, even avoided the backyard privy, encumbered with … awkward clothing and aware of watchful neighbours' (R. Anderson 2008, 8). White, bourgeois, feminine respectability was at odds with elimination. Just as the Victorian lady was said to be uninterested in sex, she was also said to be embarrassed by the sight of the toilet.

If we are to take Victorian pornography as an indicator of urinary activities, it appears that women did 'relieve themselves in the gutter' (R. Anderson 2008, 7). But it was more often the case that ladies delayed elimination to the point of anguish. Havelock Ellis (1940) wrote about an encounter with a female nurse wheeling a baby in a perambulator, and while the 'nurse stood still ... [he] heard a mysterious sound as of a stream of water descending to the earth' (quoted in Cooper, Law, and Malthus 2000, 421). The dress provided a cloak of privacy to disguise what Ellis referred to as a mysterious emission. The where and how of excretion was mediated by gender. 'It would have been reasonably easy for women to urinate inconspicuously in public spaces if necessary ... Although the fashionable skirt narrowed in shape at the end of the 1860s, restricting movement of the legs, fewer undergarments were worn and drawers remained open between the legs' (ibid., 421). Cooper, Law, and Malthus offer an interesting discussion of early urinettes built in London for women. 'They were like a small water closet but required less water for flushing. A curtain rather than doors provided privacy ... Evidently urinating, as for men, required a less stringent level of privacy than other functions' (425). People did, however, complain about the 'improper use' of the urinette, as when women defecated in a receptacle that was meant to contain urine exclusively. It was allegedly the 'poorer classes of women' who shat in the urinette (ibid., 425), indicating that the body was disciplined by ideas about femininity and class[17] as well as about sexuality. Men were also accused of indiscriminate or 'promiscuous' urination in public. Havelock Ellis (1929) noted that young men and boys could be seen peeing on city streets: 'Men were expected to be furtive and discreet, but officials understood that they would urinate *alfresco* with comparative frequency' (R. Anderson 2008, 7). Records collected by city planners in Toronto similarly suggest that indiscriminate excretion by men was common but largely ignored by police because it was widely known that there were not enough toiletry provisions for the populace (ibid.).

The spectacle of excreta on city streets in eighteenth-century London, Paris, and other western-European and North American cities came to be associated with disease (the transmission of which is meant to be curtailed by public lavatories), contagion, and death. Urinary and faecal deposits in city streets were health and safety issues, but today their visibility upsets reigning discourses of individuality. In her discussion of bodily decay, incontinence, and the social organization of care work in the present-day context, Isaksen (2002) notes that

> Having status as an individual, separate person and being accepted as a social being seems to be (at least in Western cultures) linked to ideas of not letting one's organic body functions be visible and present in social

interaction ... [we] protect ourselves by putting up affective walls ... [we commit to] social mechanisms that might have the denial of bodily decay and death as an end product. (793)

Body residues upset our psychically invested and imagined defences against death; the reigning discourse of individuality and subject autonomy being only one such defence. Gender identity is, as I argue in subsequent chapters, a comparable, psychically invested defence against subject dissolution. 'The fear of death ... is not simply about the collapse of the subject, but also about the breakdown of the body' (Wilton 1998, 177). Georges Bataille (1957), affectionately named the scatological theorist, writes that the 'horror we feel at the thought of a corpse is akin to the feeling we have at human excreta' (57).

Gender-segregated toilets emerged as London's sanitary engineers, inventors, and city planners were made aware of the threat of disease (cholera in particular) and how it was spread by sewage contamination of water intended for drink. There is a conceptual equation between the gendering and privatization of lavatory facilities and the need to clean up excreta in European and British city streets. With sanitation reform, a more private, bounded, gendered body was produced in city space. Purified by the inscriptions 'Ladies' and 'Gentlemen' and channelled into separate lavatory closets and compartments, the body was, in effect, subject to quarantine by gender. Gender purity, its intelligibility and segregation by type of genitals, was associated with health and well-being – longevity, sanitation, and protection from disease. As Žižek (1997) argues in *The Plague of Fantasies*, 'death [not gender] is the symbolic order.' What appears to be about gender may be more accurately read as a worry about human mortality. While most of us do not die or contract disease and sexually transmitted infections in public restrooms, there is, to use Butlerian parlance, a contemporary identitarian worry about coming undone. Gender identity, as Butler (1990, 2004) argues, is adopted as a psychic response to loss and prohibition. It has also been represented as a shield or defence against death and decay (Ian 2001, 82).

In a fascinating discussion of the history of shit and its relation to capitalism, Dominique Laporte (2000) notes that for the 'hygienists, shit was the site of the irredeemable, even incommensurable loss, which they were obstinately bent on denying' (124). The advent of the water closet was spurred on, if not over-determined, by the historical spectacle of death. 'Ladies' fell into cesspools and suffocated in their own faeces. They also fell through garderobes (loos in medieval castles) 'projected from the house over the space between the side wall and that of the neighboring house, so that the pile grew up neatly, out of the way of the front door but still accessible for clearing out from the street' (McLaughlin 1971, 31). Children fell into London's city sewers and drowned. 'John Michie, the proprietor of the Michie Tavern in Virginia in 1784, complained of the necessity of rescuing drunk visitors from the outhouse after they had fallen into the privy hole' (Horan 1996, 70–1). Plumbers also drowned while trying to perfect the flush-valve systems in water closets; explosions while they were working with the new technology were not uncommon. In fact, the building of city-wide sewers and public lavatory facilities in London and Paris and other European cities from the nineteenth century onward was driven by the fear of death and disease. Plumbing is overdetermined by worries about human mortality.[18]

The pairing is alive and well and acted out in the present day. For example, public toilets are, not infrequently, the site of hate crimes – physical and sexual assaults. They are venues where parents worry about children being abducted and subject to paedophilic attacks. Lavatories are also places where syringes are left behind and where people worry about contracting sexually transmitted diseases such as HIV/AIDs. People vomit and get sick in restrooms, ingest medications and illegal substances, or monitor stool for blood and observe its weight, consistency, shape, and colour as a general indicator of health.

There is a metonymic relationship between disease, psychic loss (as theorized by Butler in her writings on gender/melancholia), and gender purity. The triangulation is mapped onto the toilet. In his discussion of nineteenth-century sewage-treatment centres in London and Paris, David Pike (2005) notes that sewers 'accumulate waste, not only excrement and offal, but the cast-off and outmoded remains of things, places, people, techniques, and ideas for which physical and conceptual space no longer exists in the world above' (52). The lavatory is a museum or relic of the past. It is a storehouse for what has been lost and foreclosed in the making of the modern gendered and sexual body.

> The underground fascinates not merely because it contains all that is forbidden, but because it contains it as an unimaginably rich, albeit inchoate and intoxicating, brew of other times, places, and modes of being in the world, and because that brew intimates the fragility of the unity claimed by the world above. (Ibid., 56)

As Pike suggests, the genesis of the sewer parallels the maturation of the white bourgeois subject. As the Victorians tended to banish talk about the body and its 'lower' extremities to innuendo and euphemism in polite discourse, they also linked the genitals – their orifices, fluids, and excretory mechanisms – to the underground (the bowels of the city) and to those racialized and classed as immoral, degenerate, and/or vagrant. The city sewer was linked to the body's genitalia and abject fluids: 'The body and its continence, which modeled the boundaries of the middle-class individual self, could only be preserved through a careful policing of the abject and the closure of the boundaries of the body, through which contaminated or contaminating fluids should neither enter nor escape' (Gilbert 2005, 79). Cleaning up the sewers was also a way of sanitizing the public (or social) body. The poor, the criminal, and the prostitute, along with the racialized 'degenerate,' were all linked to raw sewage: all were in need of evacuation from 'above-ground' pedestrian streets.

The hygienic superego of today is concerned about gender purity and the eradication of all that is deemed anomalous, messy, unintelligible, or out of place in white heteronormative and cissexist cultures. Some semblance of corporeal order is maintained by securing the genitals and consolidating the form, shape, and contours of the body, including its skin, orifices, and erotic zones, by gender. Body fluids (culturally coded as abject) disturb gender identities and fantasies of subject coherence. Subject integrity is often felt to be compromised in the bathroom because it is a place where orifices open and fluids are expelled (lost).[19] Without overstating the case, people sometimes feel these openings, urinary and faecal remains, along with the moratorium on anal eroticism, to be matters of life and death.

An obsessive interest in cleanliness may, in fact, be read as angst about gender integrity and its inability to protect us from disintegration (psychic and corporeal). If gender is a defence against loss and also a way of coming undone (Butler 2004), then the spaces in which we feel our genders to be precarious or insecure will also be rooms where some will be moved to panic.

Foucault reminds us that modern disciplines were dependent upon the plague; upon the idea of the leper, and his capacity to infect. The medical lab, like the lavatory, is saturated by bright fluorescent lights – much like the scientific lab in which disease, germs, and viral infections are made visible. The 'pedagogy of examination,' perfected in the lab and upon the medical diagnostic bed and operating table, is now commonplace in present-day disciplinary institutions. Stallybrass and White (2007) observe the same pedagogy of examination and surveillance in their discussion of London sanitation reform and the management of the 'Great Unwashed' (an expression designating the poor and criminal classes in England).[20]

The public lavatory, as a modern-day gendered institution, embraces the pedagogy of examination analysed by Foucault. The body is, in this space, an object of visual inspection. Gaps between the perceived sex of the body, gender identity, and the insignia on toilet doors are subject to inquiry. The space is designed to authorize an invasive and persecutory gaze. Mirrors, fluorescent lighting, and metallic surfaces all invite voyeuristic attention. Like that of an Olympic athlete urinating into a cup in submission to a controversial (and highly dubious) sex test,[21] our gender is subject to survey every time we enter the lavatory. Prying eyes attend to the body and whether or not it is in the 'right place.'

If, as I suggest above, the lavatory is a memorial to what has been lost or foreclosed in the making of the self, it stands to reason that it acts as a cultural repository of the unconscious. It is often forgotten that the histories of sexuality authored by Foucault (1978) do not negate the unconscious. If anything, *The History of Sexuality, Volume I*, may be read as a commentary on the social unconscious: the use of the law by the masses to police and enjoy what has been foreclosed. In this volume, Foucault explicitly recognizes the role of psychoanalysis in both the deployment of sexuality and the undoing of the effects of repression upon the individual (129–31). A Foucauldian study of sexuality, focusing as it does upon discourse, power, and perverse pleasure, should not be seen as incompatible with what Judith Butler (1997b) calls the 'psychic life of power.' Gender, for Butler (1990 and 1997b), is about loss and social taboo. If sexuality is the linchpin joining the body to social configurations of power, as Foucault claims, then the psyche is necessarily implicated in disciplinary power. We cannot fully appreciate emotional investments in gender without a theory of the psyche. Nor can we pretend that geographies of exclusion do not impinge upon the emotional lives of individuals in ways that are less than transparent. What it means to be told that one is 'in the wrong bathroom,' to be sexually propositioned before a urinal, to be engulfed by the aroma of another in a neighbouring stall, to cry or to bond with a friend in crisis, or to be excluded from bathroom sociality altogether is, strangely – or perhaps queerly – missing from present-day academic inquiries.

Gender and the Bodily Ego

In order to understand the centrality of gender to the cultural politics of the present-day North American toilet and how those who are LGBTI are vulnerable to excommunication (and to gender identificatory injuries), it is important to understand the bodily ego. Trans studies and psychoanalytically informed feminist theories of the body are all, in different ways, using the concept of the bodily ego to understand gender identity and embodiment (Butler 1993; Prosser 1998a; Salmon 2004; Silverman 1996). For example, in *Second Skins: The Body Narratives of Transsexuality*, Jay Prosser (1998a) offers an important discussion of transsexual embodiment focusing on what psychoanalysts, following Freud, call the bodily ego.[22] He claims that what drives the transsexual trajectory is a feeling of 'bodily alienation ... a discomfort with their skin or bodily encasing: being trapped in the wrong body is figured as being in the wrong, or an extra, or a second skin, and transsexuality is expressed as the desire to shed or to step out of this skin' (68). Of course, not all trans people transition in a linear fashion and/or feel their skins and bodily encasings to be trappings. Some folks may negotiate very different kinds of transitions that cannot be charted along a unidirectional compass (male to female, female to male). Transitions are attempts to bring the material body into alignment with gender identities. There are various ways through which transitions are negotiated, such as using hormones (testosterone or estrogen), undergoing sex-reassignment surgeries, and changes in gender pronouns, legal documents/identification papers, and names. Some people prefer to invest the body, its genitals, contours, and shapes, with significations that are not clearly masculine *or* feminine but sometimes a combination thereof. In all cases the bodily ego, regardless of one's gender identity or trans status, is simultaneously sensorial (or material) and psychically invested.

While the bodily ego, with its material and sensorial capacities, is an important area of investigation (now central to trans studies), there is a growing agreement that, as Patricia Elliot (2001) argues, the bodily ego is shaped by processes of identification, signification, and loss. Elliot suggests that a willingness to avow and to refuse body parts (there and not there for others to see) might productively be understood to involve psychic structures. To feel 'at home' in one's body, to assume the contours of one's sexed body, it is, as Elliot suggests, necessary to find ways to deal with loss and trauma and to 'give up the "romanticized ideal of home"' (306). This is true both for those who are cissexual and for those who are transsexual.

If the bodily ego is forged in the place of or at the expense of a romantic or nostalgic image of home (necessarily lost in the course of maturation), it makes sense to consider how architectural designs are built upon white heteronormative and cissexist fantasies of home. If the toilet is a private oasis in communal space, a field of intimacy in public, it might also be thought of as a homely room where we project ourselves onto otherwise common space. How the room comes to be saturated with gender-normative imagery may be explained by recourse to the bodily ego. If the bathroom is an intermediary or tertiary space, a border, recess, cavity, or social underground where the private (body) and public (communal body) come together, it must also be a room in which definitions of self, identity, desire, and hate are negotiated (sometimes violently and with an excess of passion).

The bodily ego is a psychoanalytic term denoting the phantasmic process through which individuals learn to map, avow, embody, and give shape and consistency to a material body. For Freud, the bodily ego is a mental projection of a surface and, as such, involves a narcissistic investment in the self – as coherent – in ways that are not determined by biology. Freud writes that the ego 'is first and foremost, a bodily ego; it is not merely a surface entity, but is itself the projection of a surface' (Freud 1961 [1923], 27). In other words, the parameters of the corporeal body (its skin, material coordinates, and contours) do not define or delimit the bodily ego. The bodily ego is the result of a libidinally invested projection. In an important article about the bodily ego and transsexuality, Gayle Salamon (2004) explains that 'the *body can and does exceed the confines of its own skin* ... both body and psyche are characterized by their lability rather than their ability to contain' (108). Knowing where one's own body ends and another body begins is central to subjectivity. The bodily ego enables the subject to locate the self in space. Libidinal zones, orifices, and points of contact between the self and others are all central to the formation of a bodily imago through which the bodily ego is concretized (Schilder 1950, referenced in Silverman 1996). Through mental projections and sensory systems we can ascertain the shape, configuration, and contours of a gendered body.

Judith Butler (1993) contends that gendered morphology is mediated by a heterosexist symbolic that incites a fictitious ('imaginary' in the Lacanian parlance) certainty about the co-determinate relationship among the body, its genital apparatus, gender identity, and sex. Salamon

(2004) notes that bodily egos are not transparent reflections or reliable mappings of the body but 'allow for a resignification of materiality' (117) precisely because they do not correspond to corporeal coordinates. Some bodily egos may depart from material coordinates more than others (and this is certainly the case for transsexuals), but regardless of a person's status as trans or cissexual, everyone must negotiate a disjuncture between an internalized body image and the external contours of the body. By drawing attention to this gap, fissure, or lack of alignment between the image and the corporeal, Salamon suggests that we may challenge presumptions about an illusionary wholeness upon which people typically understand the body and its relation to gender identity. Appealing to psychoanalytic theory, Salamon notes that subjectivity is 'fragmentary and incomplete, comprised of a body and a psyche. Not only do these two elements not add up to a "coherent" whole, neither body nor psyche can be properly thought as whole or complete' (104). Without an illusionary whole, it is difficult to naturalize conventional sex and gender systems. Heteronormative coordinates and grids upon which cissexual gender identities are tied to a so-called given morphological body are unstable. The 'lesbian phallus' is, of course, Butler's key example of a psychically invested body part that doesn't correspond to a visible anatomy, though it is somatized through what we might call a transmasculine, butch, and/or lesbian identification.

Heteronormative grids cannot predict or mandate the structure of the bodily ego. While the eroticization of the body, its skin, fluids, genitals, and so on, is mediated by heterosexual matrices, the imaginary schemata (which is image-based for Lacan) never entirely assimilates to its prescribed coordinates: 'That is, the body that one feels oneself to have is not necessarily the same body that is delimited by its exterior contours, and this is the case even for any normatively gendered subject' (Salamon 2004, 96). Likewise, our genders and sexualities do not necessarily fit the libidinal coordinates of a cissexist and heterosexual order. As Butler (1993) suggests, we need a 'displacement of the hegemonic symbolic of (heterosexist) sexual difference and the critical release of alternative imaginary schemas for constituting sites of erotogenic pleasure' (91). We also need alternative geographies in which transsexual and/or genderqueer spatialities of the bodily ego can be mapped onto the public domain. Rather than building rooms in public to fit and to promote rigid gender binaries predicated on an illusionary wholeness, and policing these gender-specific divisions of space (along with those who may enter into them), we must consider how architectures may be designed to incorporate alternative images of gender that are not transphobic and exclusive to cissexuals, thereby making room for modes of identification at odds with heteronormative and cissexist landscapes.

If the bathroom is a place where gender is mandated, it is also a place where the precariousness and fragmentation of the gendered bodily ego is felt. As I discuss in more detail in chapter 3, there is an important visual component to the bodily ego that depends upon what Lacan calls 'misrecognition' (*mis*recognition because the ego comes to recognize itself through a specular image in the mirror – an externalizing and hence fictive identification with a visual mirage). According to Lacan, there is an externalizing element to the mirror stage because the image is outside the contours of the body. The subject, therefore, identifies with the image that is not the self proper but a reflection. 'Lacan suggests that the subject's corporeal reflection constitutes the limit or boundary within which identification may occur' (Silverman 1996, 11). The image enables the subject to project boundaries onto a previously undifferentiated infantile body space. Lacan describes the psychically invested contours of the body as 'an identification with a form conceived of as a limit, or a sack: a sack of skin' (Lacan quoted in Silverman 1996, 11). It is usually the case that the ego forges identifications that are relatively consistent with the corporeal imago. One does not typically lay claim to a genital structure or corporeal form that is not visualized through a glass mirror or through other people who stand in as mirrors. Gender identifications are usually delimited by the image in the mirror – along with culturally specific sex and gender systems that authorize dimorphic gender imagery. But the bodily ego is not derived exclusively through visual economies. The bodily ego is shaped by physical sensations and erotic zones as well. Sexuality is significant. Through sexual acts, fantasies, and desires (acted upon and not), we position ourselves (and are positioned by others) in ways that affirm and upset whom we take ourselves to be.

Paul Schilder (1950), a Viennese psychoanalyst who wrote about the corporeal ego in *The Image and the Appearance of the Human Body*, made important observations about the role of the orifices, the body's points of entry and exit, in ego formation. Similarly, Freud based his model of Oedipal development upon bodily openings, oral, anal, and genital. The investments placed in each region are said to shape adult gender and sexual identities for both the child and the adult. If the gendered bodily ego is negotiated through a range of sensory systems beyond the visual (including acoustic, olfactory, and tactile sensations), we may better understand why the bathroom – as a geography containing stray fluids, sounds, and smells – is felt to upset gender and subject integrity. Bodily egos are sensorial, as Freud tells us, and are shaped by acoustic, tactile, and olfactory envelopes (Anzieu, 1989): sensory systems through which we cultivate a seal (or what Didier Anzieu refers to as a 'skin ego') or bodily boundary. Silverman (1996) tells us that there is a 'nonvisual mapping of the body's form' (16) in space that should, ideally, corroborate an internalized visual mirage of the body. People strive towards a 'smooth integration of the visual imago with the proprioceptive or sensational ego' (ibid., 17) so that they can arrive at what Lacan terms *méconnaissance*: the misrecognition foundational to identity that produces an internal bodily coherence. A lack of integration, by contrast, produces a felt lack of corporeal integrity. Transgender subjectivity is marked by an explicit 'lack of integration,' but as both Silverman (1996) and Salamon (2004) agree, the disintegration is part and parcel of human subjectivity. The gender embodiments had by those who are trans and cissexual should be characterized not by an absolute difference in the way the visual imago and the sensational ego are psychically negotiated but rather by a difference in the degree to which each is felt to be compatible. Elaborating upon French psychoanalyst Henri Wallon's (1934) analytic of the visual imago, Silverman (1996) explains that the ego is *'always* initially disjunctive with the visual image, and that a unified bodily ego comes into existence only as the result of a laborious stitching together of disparate parts' (17). The visual imago and the sensational body are always difficult to coordinate, and the extent to which we bring them into alignment is always precarious and temporal.

As I suggest in later chapters, bathroom encounters shatter illusions about the coherence and stability of the bodily ego. They call upon us to imagine the self in 'bits and pieces.' The bathroom highlights the extent to which one loses the self: 'the body schema is continually losing certain elements, such as excrement, fingernails, and hair, which afterwards still remain in a psychological relation to the body' (Silverman 1996, 21). How we might live the 'heterogeneity of the corporeal ego' (ibid.) (and consequently evade narcissistic and destructive identifications predicated upon absolute incorporation and excorporation) is the central and important question. In bathrooms, trans folk are impelled to assume the corporeal coordinates of what Silverman calls an 'abhorrent visual imago' (ibid., 29). Meanwhile, the non-trans subject can – for just a moment – cultivate an imaginary, internal coherence between a libidinally invested gender identification and his or her own bodily coordinates by disidentifying with the gender non-normative subject. In other words, the fantasy of a stable cissexual bodily ego is purchased by deriding trans people, who must very often negotiate a wider gulf between the visual imago and the sensational body than the conventionally gendered cissexual subject. A similar process of denigration is at work in both structures of racialization, where, in the terms offered up by Frantz Fanon (1986) in *Black Skin, White Masks*, the black male subject is forced to identify with a white man's mirror, and in the case study of disability, where the physically disabled are imposed upon to identify with the able-bodied.

In her discussion of morphological difference and the Lacanian mirror stage, Margrit Shildrick (2009) notes that those with disabilities (like those without disabilities) are, in the psychoanalytic model, compelled to introject an idealized, able-bodied image (mirrored by the non-disabled other) in order to internalize a cohesive and uniform self-image. But this introjection cannot be uncomplicated. Those with visible physical disabilities must, she surmises, be 'radically shaken by ... [the] mark of dis-unity in the external image' (Shildrick 2009, 121). She suggests that the disabled body is 'insufficient as an object of desire' (ibid.) and is deemed unwhole in able-bodied landscapes. In this way, the disabled person, not unlike people of colour denigrated before the white man/woman's mirror image, 'becomes other, its self-positioning as a subject of desire – like that of women – denied recognition' (ibid.). Those compelled to live with the burden of difference from a normative ideal image are imposed upon to work creatively with what José Esteban Muñoz (1999) calls disidentification: a desire for a normative ideal but with a difference marked by a simultaneous disavowal and reconfiguration of that same ideal. As summarized by Kaja Silverman (1996), there is a 'psychic dilemma faced by the subject when obliged to identify with an image which provides neither idealization nor pleasure, and which is inimical to the formation of a "coherent" identity' (27).

The most salient examples of incongruity are evident when the disabled are degendered by the signage on the bathroom door (toilets for persons with disabilities are often gender neutral); when women of colour are seen before a mirror (and against an ultra-white and sanitized backdrop) by those who more closely approximate a white, Euro-American ideal of feminine beauty; when trans folk are subject to visual surveillance in the stall, before a urinal, and in the common floor space before the mirror by those with cissexist privilege, and so on. Mirrors are endemic to the toilet and deployed to monitor sameness and difference, self and other, in ways that bear upon bodily ego functioning. The scene of recognition before the mirror is sometimes brutal because it decides who will and will not be recognized as a normative subject. 'No matter how much we each desire recognition and require it, we are not therefore the same as the other, and not everything counts as recognition in the same way' (Butler 2005, 33).

In his seminal article on space, the psyche, and landscapes of exclusion, Wilton (1998) notes that changes in the 'morphology of social space may be experienced at the level of the bodily ego' (176). In other words, social subjectivity is spatialized. Spatial configurations also tell us much about systems of exclusion predicated upon a need for gender and bodily integrity. The toilet amplifies the differences and points of interconnectivity between what Silverman (1996) calls the 'self-same body' (11) (cultivated by the conventionally gendered and/or heteronormative subject) and the 'de-idealized body' (20) (denigrated *as* queer, trans, or disabled, or racialized as non-white and in other ways non-normative). The 'self-same body' is dependent upon an aggressive identificatory structure predicated upon racist, ableist, classist, cissexist, and heterosexual matrices. As Silverman explains, 'the ego consolidates itself by assimilating the corporeal coordinates of the other to its own – by devouring bodily otherness. The "coherent" ego subsequently maintains itself by repudiating whatever it cannot swallow – by refusing to live in and through alien corporealities' (24). The 'de-idealized body' is subject to repulsion and expulsion because it is culturally coded as abject.[23] In this way, the bodies of others are either incorporated (aggressively assimilated) or abjected (ejected or defensively refused). Our relations to others are predicated upon a continual assessment of sameness and difference. Both incorporation and expulsion are central to bathroom sociality. We either assimilate or expel others – those who do and do not consolidate our libidinally cathected body maps.

Perhaps Judith Butler (1990) said it best in her discussion of gender, identity, and abjection when she wrote that there is a 'mode by which Others become shit' (134) in object relations. Butler contends that the normative subject boundaries between interiority and exteriority are 'confounded by those excremental passages in which the inner effectively becomes outer, and this excreting function becomes, as it were, the model by which other forms of identity-differentiation are accomplished' (133–4). She notes in a later publication that abjection 'designates a degraded or cast out status within the terms of sociality' (1993,

243, note 2). In other words, social abjection denotes a process through which people constituted as Other, or as different, are literally banished to 'zones of inhabitability' (Butler 1990, 243) or used as objects of projective identification. In the case of projective identification, what Calvin Thomas (2008) calls 'scatontological anxiety' (69), 'the fear of being abjected, of *being something not worth having*' (Thomas 2008, 70), or, as I suggest in this study, the fear of being a subject who doesn't matter, or, alternatively, one without a urinary door or sign to authenticate entry, is transferred onto LGBTI people: 'Abjection assuages, discharges, or "gets rid" of a subject's own "god-awful feeling" of scatontological anxiety by punitively projecting that affect onto a degraded "other" who is forced to assume the fecal position' (Thomas 2008, 147).

Silverman (1996) also notices the 'refusal on the part of the normative subject to form an imaginary alignment with images which remain manifestly detached from his or her sensational body, and his or her stubborn clinging to those images which can be most easily incorporated' (24). People tend to aggressively assimilate the corporeal coordinates of the other or refuse the difference of the other altogether in callous disregard and disassociation. Both are equally violent in their erasure and negation of difference. The ego is continually monitoring itself and others for images that resemble and differ from the self. While subjects are invested in mirror images, the bodily ego has no essential or stable point of reference in space. As Silverman notes, the bodily ego 'undergoes repeated disintegration and transformation' (13). This experience of 'disintegration' and 'transformation' is unnerving. Gender needs to be consolidated by what Judith Butler (1990) calls a psychically invested gender performativity. One way to consolidate normative gender identities has been to order bodies in toilets by gender and genitals. This book questions the ethics of this ordering and invites us to think differently about how we position ourselves and others in space.

2 Trans Subjects and Gender Misreadings in the Toilet

The gender-segregated public water closets of the Victorian era are now institutions in the present-day Canadian and American context. Public toilets in transit stations, malls, shopping centres, gas stations, sports arenas, concert halls, workplaces, schools, colleges and universities, restaurants, and bars are all – as interviewees explain – venues in which gender is subject to contestation and debate. Judith Halberstam (1997b), who has written about the 'bathroom problem,' confirms that it is 'no accident that [public] travel hubs become zones of intense scrutiny and observation' (177) because we often feel a need to secure identities when our bodies are in motion or transit. While public lavatories were intended to 'clean up' city streets in the nineteenth century, to make the excretory habits of the poor and immoral visible and subject to hygienic surveillance (Stallybrass and White 2007), they now, in the contemporary North American context, function to discipline gender in ways that are mediated by class, 'race,' and visible physical disabilities. Those who are trans and/or perceived to be gender non-conforming are subject to visual and verbal scrutiny in public toilets.[1]

The architectural design and gendered codes of conduct mandated in the lavatory all support the illusion that there are two binary genders – male and female – both of which are visible, identifiable, and natural.[2] Hygienic imaginations, originally focused upon disease (cholera and typhoid in particular), inspection, and quarantine in the nineteenth century, are now focused upon whiteness and gender purity. 'Cleaning up,' or rather 'sorting out,' the differences between male and female, masculine and feminine is a function of today's gender-segregated toilets. But little attention has been devoted to the effects of gender-segregated facilities on trans people, who are especially vulnerable to excommunication in gendered public space.

When gender is mistaken – subject to moral and aggressive scrutiny – people are, as Judith Butler (2004) suggests, 'undone.' To have one's gender identity questioned and interrogated is to have one's desire for recognition provisionally foreclosed. 'If part of what desire wants is to gain recognition, then gender, insofar as it is animated by desire, will want recognition as well ... recognition becomes a site of power by which the human is differentially produced' (2). The difficult question of who gets to count as a gender-normative subject, under what conditions, through whose estimate, and in what social space, is about quality of life. Gender recognitions consistent with self-identifications are not inconsequential or superfluous but gateways to humanity. For those who are routinely denied access to public space, humane gender recognition is known to be essential to community membership.[3]

When public space is rigidly gendered, access will be an issue for those who do not conform to the norms upon which sexual difference is consolidated. When one is denied access to public space, questions must be asked about the relationship between gender recognition and erasure, entitlement to public participation, and access to personhood. As Temperance, who is a white Anglo-Saxon Protestant (WASP), a non-trans queer femme, and a graduate student living in Toronto, astutely notes,

> For trans people who ... don't identify their gender or just know it's not one of the two [binary gender] extremes ... I could imagine that ... [accessing a public bathroom] is the most triggering event of their day. To just, like, have to pee and have to be confronted with the fact that you can't go, you literally don't have access, you don't have a door. It's almost like you are entirely erased from the most human and basic and fundamental of activities.

Not having a door (or a sign) is a pertinent metaphor for those who have their gender identities rendered invisible, subject to erasure, or expunged from the social field. To be unseen, to be unrecognizable,[4] to be interrogated by indignant onlookers upset about gender 'impurity' or incoherence, and to lack legibility in a cissexist and heteronormative landscape is to have one's legitimate access to public participation thrown into question.

Gender misreadings are points of anxiety and contention in the public lavatory. Using interview testimonials from trans people, I demonstrate how there is a late-twentieth- and early-twenty-first-century cissexual worry about a discord between gender identity and sex. Trans people do not always have access to what Serano (2007) calls cissexual privilege.

Cissexual privilege is typically given to those who are not trans and thus more able to orchestrate a normative concord between their gender identities and the sex of their bodies, as perceived by others. Of course, some trans folk can (and do) access cissexual privilege. Those who have their gender identities read 'correctly' in the normative landscape (or those who can use their bodies to authorize a claim to a given gender identity) access cissexual privilege. To have one's gender identity questioned is to lose cissexual privilege and, consequently, to face what Serano (2007) calls cissexism, the 'belief that transsexuals' identified genders are inferior to, or less authentic than, those of cissexuals (i.e., people who are not transsexual and who have only ever experienced their subconscious and physical sexes as being aligned)' (12). To have one's gender identity questioned may even incite physical assault and verbal interrogation or police arrest and removal by security guards (to be discussed below). But to have one's gender identity questioned is also to be shamed and ostracized in the public eye. Part of what it means to come undone is to be effaced or rendered invisible (Namaste 2000). As Susan Stryker (2006) notes in her discussion of Frankenstein and transgender rage, 'Transsexual embodiment, like the embodiment of the monster, places its subject in an unassimilable, antagonistic, queer relationship to a Nature in which it must nevertheless exist' (248).

By invalidating the gender identity of another, people commit a psychic act of violence, foreclosing upon the subject whose gender identity is misread. To be a subject is to be recognized in a social field in which gender matters. Gender is a linguistic and cultural currency of sorts. Injuries to person occur through repeated and intentional misgenderings (Serano 2007, 179). Susan Stryker (2006) further notes that we are situated in a 'field governed by the unstable but indissoluble relationship between language and materiality, a situation in which language organizes and brings into signification matter that simultaneously eludes definitive representation and demands its own perpetual rearticulation in symbolic terms' (252). Subjective space is foreclosed when folk defy or live in opposition to these terms, even when that foreclosure is 'grounds for the materialization of ... [bodies] and ... [the] bodily ego' (ibid., 253). Trans subjectivities are often forged in the gaps, the vacuous spaces between exacting and exclusive male and female signs. The place of annulment is trans-generative as it is also injurious. This chapter maps out the injuries to self incurred by gender-exclusionary designs and offers insight into how binary gender categories – their logic, purity, and intelligibility in cissexual imaginations – can be subject to resignification.

Gendering Bathrooms

Interviewees talk about a variety of public reactions to their presence in the toilet, ranging from curiosity, inquisition, surprise, confusion, and avoidance to fear, anger, hostility, and hatred. Rachel, who is queer and butch, says with frustration,

> I get harassed every time I go into a bathroom, whether it's, like … a woman jumping back as I walk in the door, or people giving me dirty looks, to, like, full-on confrontations in bathrooms … I've had a security guard ask me to leave … a drunk girl at the Madison [a Toronto bar] start screaming, 'There was a boy in the washroom.' So I just get my back up every time I go into the bathroom.

Rohan, of Anglo-Celtic heritage, trans, butch, queer, and living in Toronto, explains how a masculine presence in the 'women's' toilet leads to troubling looks and verbal interrogation:

> The majority of the people in [the 'women's bathroom'] … either mistake me for a man or are deeply troubled by the presence of a masculine person in a woman's washroom … it's always a problem. Always … I get strange looks, comments … being interrogated about whether or not I should be in there … stared at a lot, spoken about as if I am not there … some people are just nervous and some people are just not sure [about my gender].

Ivan, a Portuguese-Canadian, non-trans man, queer, and also living in Toronto, notes that

> Trans folk, people who are on the continuum of gender identity, and who are not easily recognized as one or the other, like these liminal zones, are the ones who get the most trouble or harassment in our culture. Our culture hates things that are ambiguous … People are threatened by ambiguity. People are threatened by boundary crossings. So, people who don't fit, people who are not immediately recognizable as one gender or another … are obviously going to receive the most scrutiny, the most barriers, prejudice, perhaps active blockage from using the bathrooms.

The vast majority of interviewees say that the gendering of toilets has to do with heteronormativity along with a binary gender system intolerant of those who lack access to cissexual privilege. For example, Butch Coriander, who is a WASP, non-trans woman, genderqueer, butch-dyke, a professor, and a community activist living in Toronto, laments, 'We are so policed around our gender ... and it's almost like it's so fragile that it needs to be reinforced constantly and probably that's because so many people don't fit.' As David, who is of European heritage, Christian, a non-trans man (who identifies as androgynous), gay, and living in Kingston, Ontario, similarly notes, 'I think that people in mainstream society are *really, really* strongly attached to the idea that there are men and ... women and anything that doesn't fit into one of those boxes or in this case, rooms, just is an aberration.'

Those who are seen as trans and/or as gender non-conforming upset, and are seen to be in defiance of, gender signs on bathroom doors. They consequently upset heterosexual matrices (Butler 1990) stabilized by cissexist and mutually exclusive gender signage. Sexual difference is often thrown into question for cissexual patrons who, upon encountering gender difference, are forced to question the 'nature' of gender in ways that are at odds with conventional logic. Emily, who is WASP, intersex (surgically altered at birth and given a male gender identity by doctors, and transitioned to a female gender identity in adulthood), lesbian, living in Kitchener, Ontario, who regularly uses the 'women's' toilet, says that 'If we don't segregate the bathrooms then we're obviously saying that there are less differences between men and women than we thought ... that is terrifying for some people.' Temperance makes the same observation: 'If we were to desegregate gendered bathrooms we would have to have a public social conversation about the fact that ... the majority of folks fall somewhere in between the signs on the bathroom doors.'

Gender misreadings are negotiated through a visual economy of power that is dependent upon optics and what Michel Foucault (1978) calls techniques of surveillance. Those who feel proprietary about gender feel entitled to stare at those who are seen to be at odds with the clearly delineated signs on facility doors. As Ivan notes, 'I think if someone doesn't look recognizably of the same gender [symbolized on the bathroom door], the [other] person will look in their eyes, will stare at them, hold a gaze longer than you need to.' Gypsey, who is of Irish-Swedish and Scottish ancestry and living in Toronto, transgenderist[5] (living full time as female), lesbian, with a visible physical disability, and a regular user of the 'women's' bathroom, confirms this view: 'I think when I walk into a room where it is known that I am transgendered, I am the most watched person in that room. What I do, which washroom I use. I think that I am always being judged by somebody else's standards.' Jacq, who is of British ancestry and living in Vancouver, genderqueer and butch, lesbian, and with a visible physical disability, uses both the 'men's' and 'women's' toilets, and confides that 'There's never a time where I go into a bathroom where I'm not looked at, given a dirty look, or something happens, like, there's a cleaner outside that says, "You're not allowed to go in there," or something. And ... it's not like it just happens once ... it's *every* time.' As Claude, of British and French ancestry and living in London, Ontario, a non-trans butch lesbian who uses the 'women's' restroom, notes, 'I definitely have gender anxiety when I go to use the toilets ... I don't look so butch that it carries on [longer than a moment] ... it's usually a sort of flicker. I get looked at and then they realize I'm not a male, but I always get looked at. There's always that moment ... .a visual exchange that is about gender recognition, whether or not it will be conferred.'

Some of these 'looks' are questioning and curious. At other times the looks are aggressive, proprietary, defensive, fearful, and/or indignant. It is often the case that hostile looks in toilets are legitimized by appeals to child protection. It is not uncommon for LGBTI people to be read as sexual threats to children. Zahabia, who is an Indian-Canadian, Muslim, non-trans queer femme and lesbian, shares her observations about how trans people and butches in particular are seen in the 'women's' toilet: 'I think about ... butches ... I know, I think about trans people that I know ... people feel quite unsafe [in public bathrooms], and they get ... everything from the funny looks to ... [parents] ... pulling children in close ...[and] staring a long time, saying you're in the wrong place.' She illustrates a not-so-subtle cultural association between masculine, butch, or gender-variant people in the 'women's' room and paedophilia. This positioning is projective and dissociative. By reading another person as 'perverse' or 'paedophilic' because of the person's gender identity, and simultaneously appealing (explicitly or implicitly) to the 'vulnerability', 'impressionability,' and 'sexual innocence' of children, parents and caregivers re-entrench the importance of the nuclear family, its heterosexual and reproductive mission, and the concordant threat posed by the gender- or sexually dissident subject. Gender impurity here becomes a threat to what Lee Edelman (2004) calls white reproductive futurism.

Of course, gender intolerance, does not 'protect' children; it produces a gender-phobic climate.[6] Visual exchanges in toilets are emotionally painful for trans people who loathe having their bodies intercepted by a de-idealizing gaze. Seo Cwen, who has ancestors from the United Kingdom and is living in Kanata, Ontario, a non-operative trans woman taking hormones (currently in transition), and gay, says: 'I'm afraid that if I do use the men's washroom, people might think that I'm some kind of gender mutant.' Trans interviewees often say that transphobic looks generated by aggressive and intentional gender misreadings are nullifying and devastating to one's person. Most emphasize the potency of what we might call a culturally de-idealizing and transphobic gaze. Jacq notes with fervour, 'The looks are ... what ... hurt the most ... They're ... giving you that real big evil look ... And ... that for me ... is more harmful than when they say to you, "You should be in the other bathroom." '

Humane gender recognitions depend upon the availability of trans-positive imagery, discourses, and 'ways of seeing' that are not reducible to binary gender assignments at birth and national identification papers authorizing two unalterable and mutually exclusive gender positions (male and female). Gender purity is based on a politics of excommunication that rejects those who are genderqueer and those who are trans, who, in the words of Julia Serano (2007), do not look 'cissexual-like.' For example, Eric Prete, who is mixed Ojibway and Celtic (raised Roman Catholic), a trans man, bisexual, from northern Ontario, and actively involved in trans community organizing, confirms that 'Passing [as the gender to which one identifies] is when you walk into the bathroom and nobody stares.' As Prete notes, 'passing' affords tremendous gender privilege. For this reason, the term is frequently used by trans people to indicate differing experiences had by those who are and are not subject to gender misreading. But as Serano (2007) explains,

> While the word 'pass' serves a purpose, in that it describes the very real privilege experienced by those transsexuals who receive conditional cissexual privilege when living as their identified sex, it is a highly problematic term in that it implies that the trans person is getting away with something. (176)

Serano further suggests that the pressure put on trans people to 'pass' enables those who are not trans to ignore what she calls cissexual privilege (178): not having to worry about the status or authenticity of one's gender identity. In other words, the impetus to 'pass' is incited by a binary gender code that is authenticated by cissexuals and inhospitable to those who do not 'enact naturalized states of being' (Boyd 2006, 421), thereby losing civic rights, protections, and legitimate access to public space.[7]

For many transsexuals, failure to validate one's gender identity is a more pressing issue than binary gender spatial divides. As Phoebe, who is a white trans woman taking hormones, living in Toronto, and actively involved in trans community activism, explains,

> I've spent thousands of dollars on transforming my body, through different procedures. I have spent a lot of personal, emotional, psychological energy on transitioning because I needed to do that to feel comfortable and happy as a human being. I'm going to damn well use the women's washroom. I went to a lot of trouble to use that. There's no other option for me really.

Jay, who is of Irish and Italian ancestry and living in Albany, New York, genderqueer, gay, and who uses both 'men's' and 'women's' facilities, also emphasizes how important gender recognitions are:

> You work so hard to have your gender perceived in a certain way and it's pretty important to have your gender [perceived] in a certain way and when you have to go into a bathroom ... that's all thrown into doubt. I think that's pretty threatening to your self and your soul.

It must also be said that some interviewees want to be recognized as 'trans' as opposed to 'male' or 'female' unambiguously. Such participants are resentful or upset by gender misreadings that nullify their cross-gender, trans, or genderqueer identities. While some transsexual interviewees want their gender identifications to be rendered intelligible *as* male *or* female, masculine *or* feminine, other trans interviewees – particularly those who identify as genderqueer – do not want to have any one single gender identity imposed upon their person.

Regardless of how one wants to be identified in the domain of gender, binary gender designs promoted and regulated by cissexuals pose a problem for many interviewees with respect to public participation and access to gendered space – particularly for those in transition and for trans people without access to cissexual privilege. Seo Cwen explains that during her transition she worried that people would see her as a 'cross-dressing fag looking to seduce gay men,' and that she is fearful of non-trans men because she 'grew up being teased for being queer, predominantly by guys, so I instinctively fear that.' This participant also notes that early in her transition she was 'very uncomfortable ... in the women's room. I didn't pass so well [in] early transition, and so I was afraid somebody might say something really hurtful. It never happened, but all the same, I was very afraid at first.' The effect of fear is often incited by worries about gender misreadings, being seen as 'gay' (or in Seo Cwen's case, a 'cross-dressing fag') in homophobic landscapes and the accompanying assaults to self. Violence has material *and* psychic dimensions. Those subject to repeated gender misreadings incur emotional injuries to self. As one trans guy (who chooses to remain nameless) speculates, 'I think the long-term effects [of gender misrecognition] are ... anxiety and ... dealing with public humiliation ... I've had, like, serious anxiety about ... being stared at.' Another interviewee who is also a trans man (and similarly wishes not to disclose identifying information) talks about his worry about gender-based embarrassment: 'When I wasn't sure that I did pass that was anxiety provoking, but more in terms of embarrassment than worrying about somebody doing something [physical].' Velvet Steel, who is of Danish origin and living in Vancouver, a post-operative transsexual woman who is taking hormones and is bisexual, notes how her self-worth is thrown into question: 'I don't feel [physically] threatened when going in [the bathroom]. I think it's more like my own self-worth, my own ego, my own acceptance of myself that gets put into question by other individuals.' Emily has avoided toilets for large chunks of her life:

> I've spent many a year being, unduly, I think, afraid of ... using public bathrooms. Afraid of, not necessarily ... physical threat, but sort of the emotional, social threat of being 'outed' and embarrassed and humiliated in front of people ... being stood up in front of everybody ... 'You're different and you don't belong here.'

Perhaps the most frequently cited effects of gender misreadings in gendered restroom facilities for trans people are shame, embarrassment, and anger. As a First Nations interviewee who is trans and two-spirited explains,

> I feel a real shame [and sense of] embarrassment because of the looks ... I'm getting ... [They are] like, 'What the fuck are you doing ...' Or, 'You're

a woman.' People don't even have to say it anymore ... you can tell by the looks they give you [that you don't belong]. And I'm not sure which is harder ... I've encountered such ... extensive homophobia verbally ... I can deal with that ... but ... it's the looks ... that seem to be able to penetrate you.

Shame is difficult to bear and is, in this case, an effect of a transphobic and de-idealizing gaze that is excommunicative. Occurring along multiple axes of difference, shame is an interlocking effect of transphobia, homophobia, and racism towards, in this instance, First Nations people in Canada. It is important to remember that the institution of the public water closet dating back to Victorian England and France was meant to differentiate people by class, gender, sexuality, and 'race.' The water closet was also imposed upon indigenous people whose lands were colonized by the English and the French. People of African and East Asian countries (among others) were racialized by excremental design and urinary habitus. In fact the existence of racially segregated toilets in the United States is a direct legacy of the will to quarantine and to segment bodies by 'race.'

Neil, who is Middle-Eastern and East-Indian, born in Kenya, and a recent immigrant to Canada (with citizenship), a trans guy, queer, and living in southeastern Ontario, reflects upon the toilet as a colonial technology that racializes non-white women in Kenya (where he was born). He notes the extent to which the 'European-style toilet' in African countries functions to racialize and class non-white women – particularly as it is presented as a superior option to the 'squatting toilet' (often branded 'primitive,' 'backward,' and 'retrograde'). Neil says, referring to the way women of colour are racialized and classed by European toilet designs, that the modalities of gender-based prejudice are 'interesting, because there are different types of women.' He also says that in Canadian public restrooms he never knows 'if someone is being sexist, racist, or homophobic, because I am a person of colour who is queer and trans.' Kew, who is Hong Kong Chinese-Canadian (with citizenship), genderqueer, and a dyke, also notices how racism and genderism intersect. She says, 'it's usually white people, it's women. They are always very conventionally gendered women. No one who's ever yelled [at me for being in the "wrong" bathroom did so] from a gender[ed or racialized position] I [was un]certain of.'

Gender misreadings inflected by racism and classism incite a variety of effects, including anger and shame. In order to understand how humane recognitions are integral to public participation it is important to remember that social exclusion is caused not only by physical violence and excommunication (forceful removal by security guards and police), but also by identitarian injury. People are shamed when they are unable to incite an affirming mirrorical response in the other. Eve Kosofsky Sedgwick (2003) notes that shame is caused by a refused recognition effectively rupturing the 'circuit of mirroring expressions' (36) – a circuit upon which social interaction is made possible. Gender misreading breaks the 'circuit of identity-constituting identificatory communication' that affirms one's status as a subject.

When subject to gender misreading, people may 'lose face' (or 'go red in the face'), wearing an identificatory stigma on the body *as an effect* (red or downward-pointing face). Sedgwick, building upon the foundational work of Silvan Tompkins, writes that 'shame effaces itself ... shame turns itself skin side out ... shame and self-display ... are interlinings of the same glove' (2003, 38). In her discussion of skin in the Western cultural imagination, Claudia Benthien (2002) similarly notes that 'shame' and 'skin' have common Indo-Germanic roots, and that both are used to denote a covering or covering up. Shame is about inhumane exposure; it makes visible a 'transitional space' (ibid., 100) or interface, what she refers to as a 'polarity or tension between one aspect of the superego, the idealized image that I have of myself, and the ego function of self-observation, the image of myself that I have in reality' (ibid.). Shame is thus wrought by exposure (vulnerability to surveillance), by having the body revealed for all to judge. Trans-specific forms of shame emerge when one cannot conceal gaps between gender identification and the body, its contours, genitalia, and orifices *as seen* by cissexuals.

What Susan Stryker (2006) calls transgender rage emerges when one cannot 'satisfy the norms of gendered embodiment' (253), when one is compelled to negotiate the 'fictions of "inside" and "outside" against a regime of signification/materialization whose intrinsic instability produces the rupture of subjective boundaries as one of its regular features' (252). The anger emerges when one is required to abide by a set of gender norms that negate one's status as trans, when it becomes necessary to 'take up, for the sake of one's own continued survival as a subject, a set of practices that precipitates one's exclusion from a naturalized order of existence that seeks to maintain itself as the only possible basis for being a subject' (253).

Tulip, who is of Israeli descent, living in Brooklyn, New York, a non-trans genderqueer femme, and bisexual, notices the way trans folk are read as 'out of place' in the gender-normative landscape, as anomalous beings:

> Allowing for gender-neutral bathrooms or spaces where trans folk don't feel like they are being looked at like 'freaks of nature' would mean that [transgender identifications are] ... okay and normal ... I don't think people are willing to do that ... [they] want to make queer and trans people feel uncomfortable and 'out of place' and not give them the rights that an average straight person has ... It works ... I think it is a [social] mechanism.

Rohan, similarly, says that, 'Every time you are forced to choose to use ... [one] washroom ... it's like you are governing yourself, you are regulating yourself.' Neil says that in instances where he is told he is in the 'wrong' washroom he is compelled to normalize his presence by saying that he is allowed to be there, lamenting that 'this robs me of any sort of trans-identification that I might have ... because [I have to respond] ... "no, I'm female." ' As Temperance reflects, 'I didn't make up these rules but I regulate my entire life by them.'

While there is no rationale for gender-exclusionary designs to be found in scientific studies of disease and contagion or in epidemiology, or even in dominant safety narratives (to be discussed in a later section), there is an irrational worry about the transferability or contagion of gender. Gender non-conformity and/or trans identities are, irrationally, felt to be contagious or, at the very least, disorienting to many non-trans people. Participants talk about how their presence in gendered toilet space is confusing and upsetting to cissexuals. Cissexuals who are not trans-literate expect to be able to intercept and then map the gender of any given occupant onto the dimorphic gender designs of the toilet. Such patrons sometimes panic when they spot trans folk in the toilet. Those panicked by trans people are sometimes disoriented and confused about their own gender identities and spatial coordinates. If, as Butler (1997a) suggests, an undeclared homosexuality is brought to the surface through a communicative exchange about homosexuality, then it might also be the case that an undeclared ambivalence about gender identity is brought to consciousness by visually apprehending one who is recognizably trans and/or gender non-conforming in the lavatory.

Transphobia indexes the gender-based disturbances had by the conventionally gendered non-trans subject. In other words, transphobic modes of address occur when one feels the gender norms by which he or she recognizes the self to be out of order or interrupted. The gender norms through which one comes to recognize the self in the other, and the other in the self, are troubled in the mirror image of the other. There is, in the mirrored disunity, a 'site of rupture within the horizon of normativity and implicitly [a] call for the institution of new norms, putting into question the givenness of the prevailing normative horizon. The normative horizon within which I see the other or, indeed, within which the other sees and listens and knows and recognizes is also subject to a critical opening' (Butler 2005, 24).

This 'opening,' made by a rupture in the mirror circuit of recognition, is sometimes felt by cissexuals to be contagious. The failure to return gender-normative imagery is intercepted by the gender-normative and non-trans subject as an invitation to (or, rather, as a demand for) self-questioning that may lead to his or her own unrecognizability as a subject. Transphobia is, as I suggest, less often incited by the presence of trans folk than by the imagined transferability of transsexuality (or gender variance). Perceptions of the other are transitive, meaning that they can involve a reversibility between the 'body's simultaneous status as perceiving subject and object of perception' (Vasseleu 1998, 29). Vision is not always objectifying (as I will discuss in more detail in chapter 5). It can sometimes function like the sense of touch, where the one touching and the one being touched are confused. While it is true that transphobic looks are often objectifying, they can also be structured by subject/object confusion. There is a not-uncommon scene in which the one confused by a trans person in the toilet identifies with the confusion, intercepted by sight, and puts the self in the place of the other who is thought to be 'out of place.' It should be remembered that the 'body of the perceiving subject is given form and content through its experience of surrounding objects' (ibid., 51). When the perceiving subject is uncertain about the body of the other intercepted by sight, the viewer may come to be uncertain about his or her own gendered body and postural schema. Merleau-Ponty (1968) suggests that vision is structured by an inherent reversibility. 'It is of the essence of visual perception that in order for me to see I must be visible for an other' (Vasseleu 1998, 52).

The reversibility of vision in the mirrorical exchange is evidenced in the following cases, where the transphobic viewer experiences the self and not the trans person coded as other (or, perhaps, not the trans person exclusively) as spatially disoriented or 'out of place.' As Jacq explains, 'Sometimes ... [a] lady will [be] looking at me and then look at the sign on the door. And then she'll pretend that she's in the wrong bathroom herself, and then she'll go to walk away, and then she'll look again, and go, "Oh that's the ladies." ' JB, who is a plumbing apprentice, genderqueer, butch, and lesbian, notices the same reversibility in the visual exchange: 'I get lots of looks ... [in the bathroom and sometimes I see] ... people backing up to check themselves, the symbol on the door, because they think that they've gone into the wrong washroom.' As a third interviewee who is genderqueer also notes, 'I have had people do a double-take where they look at me and look at the sign and look at me again and try and figure out what happened.' Rohan provides yet another account of the reversibility of the transphobic gaze and how it can disorient cissexual women:

> I am washing my hands and a woman opens the door and she sees me and she stops. She doesn't look at me she is looking at the floor. And then she steps back and she looks at the sign on the door and then she kind of just sort of looks at the ceiling and then just sort of says to the universe, '*Am I in the right place?*'[8]

The triangulation of the gaze, from self to other, other to linguistic sign (on the bathroom door), sign to self, and around again is, as indicated by the interviewees, disorienting for the patron accustomed to receiving gender-normative imagery in the mirrorical return. Public mirrors are windows into the self. The perceptual systems giving rise to gender are relational and spatial. Vasseleu (1998) suggests that the 'senses are a flesh of organic mirrors' in the work of Merleau-Ponty (1968), and that we are mutually implicated in perceptual systems, if not equally or upon a level playing field. A good example of how gender-identificatory circuits are mirrorical is narrated by Rachel. She notes that people don't always look at the gender sign on the bathroom door but at the people walking into the bathroom:

> People don't always look [at the sign] ... they look at who's going in ... [Once] I was coming out ... of one washroom, and a woman saw me come out the door and just automatically went into the other door. And so this full bio woman just walked into the men's washroom because she saw me come out of the door.

The problem with public mirror circuits is that they refract in ways that do not cohere with internalized self-portraits (what Lacan would call the visual imago). There is always something about the self that does not appear in the hanging glass or register in the optic nerve of a transphobic seer, something that defies intelligibility in transphobic landscapes.

Butler (2004) suggests that for those whose lives exceed restrictive gender norms, or for those whose gender identities are not easily seen by cissexuals unversed in the logic and cultures of trans communities, there is a question about when such departures from the norm upset and consequently undo the norm, and when the departures occasion an 'excuse or rationale for the continuing authority of the norm' (53). When and in what conditions are gender recognitions conferred, and when and in what contexts are they withheld? Equally, there are important questions to be asked about how life may be lived on the cusp of a norm for those who are genderqueer and/or trans. Speaking about lesbian and transsexual histories, Nan Alamilla Boyd (2006) summarizes the key question: 'Do abject or queer bodies retain inchoate or inherently resistant positions vis-à-vis the state? Is it necessary to transition (or pass) from abject to intelligible in order to function within the state (or in order to resist a state-sanctioned, rights-based economy of value)? How do bodies that do not matter become bodies that matter?' (421–2). How does one live a life in the absence of a mirrorical return that would otherwise validate one's internalized self-imagery and psychically invested corporeal coordinates?

Transphobic encounters in public washrooms tell us much about how people live alongside rigid and exclusive gender norms and in broken mirror circuits. Restrictive gender norms are sometimes parodied and resignified by trans people to access otherwise restrictive public space. Rachel recalls an incident in which her body was subject to a gender misreading and then reinstated into an intelligible normalizing frame:

> I was in the ['women's'] bathroom at [a bar] ... in New York City, this girl ... kind of looked at me and she goes 'Am I in the wrong place?' I was, like, 'Neither of us is in the wrong place.' I ... grabbed at my chest and was, like, 'No, no, we're all good' ... if they can notice breasts, then they're okay.

Jacq also underscores the use of 'breasts' to access the 'ladies' room: 'I used to take my coat off and stuff and my sweater off, and then I'd undo my buttons on my shirt so that part of my breasts were showing so that they [non-trans women] could actually see I had breasts.'

As Rohan explains:

> I stick my chest out, which is ridiculous because I spend the rest of my life trying to conceal it, and if I won the lottery tomorrow I would be on a plastic surgeon's table the day after and it would be gone. But it's like, it's a strategy that you are forced to undertake, even though for me that's part of my body that I don't even acknowledge let alone accentuate ... But it's ironic because ... they [breasts] don't exist for me, except for when I use the washroom.

By accentuating or laying claim to 'breasts,' butch and transmasculine folk gain entry into gender-exclusionary space. It is also a way of over-riding hegemonic equations linking 'breast/s' (dominant signifiers of femaleness) to female gender identity. Butch and trans men may have 'breasts' but invest in masculinized chests (especially after top surgeries). Conversely, trans women may have or cultivate breasts despite male gender assignments at birth. Gender performances in public lavatories, such as the ones cited above – however exasperating and tiresome to many trans folk who are forced to enact gender norms that negate their trans status and masculine gender identities – do reveal that there is something inessential about the cissexual equation between sex and gender identity. Gender identities are seen – by non-trans onlookers – to be out of sync with the way the body is intercepted in transphobic mirror circuits.

When cissexual patrons bear witness to how the rigid application of gender norms is injurious to trans people, apologies are sometimes given. It is not uncommon for people – usually cissexual women – to apologize profusely for having mistaken another person's gender identity. The apologies are said to be 'awkward,' 'uncomfortable,' 'anxious,' and 'embarrassed.' As Rohan notes, 'People are often apologetic if they mistake me for a guy and I'm, like, no, I'm in the right washroom. And they're, like, oh, I'm *very* sorry! They do feel bad. Many people do still ... feel quite bad and awkward and uncomfortable and they're very apologetic.' As Rohan's example indicates, apologies sometimes ratify gender misreading. In the above illustration, the apology is given because the non-trans patron sees Rohan *as* a woman, thereby negating a trans or masculine gender identification.

While apologies are often ambivalent and inadequate, they are (unlike the physically violent and excommunicative response) opportunities to forge positive identifications in public mirror circuits. Those uttering apologies may not always know why they are apologizing, although they may intuit that they have somehow committed an injurious (or nullifying) act. Others may give an apology to cover their own embarrassment or confusion. But to ask with sincerity, '*Am I in the right place?*' (as opposed to a rhetorical 'Are *you* in the right place?'), to utter an apology (however tiresome and inadequate it may be to those subjected to repeated gender misreadings) is to acknowledge a rupture in the otherwise gender-normative landscape. An apology may also signal a uniquely *non*-trans feeling of disorientation or confusion in gendered space. To be uncertain about one's own gender identity, particularly when one is cissexual, is to lose an epistemological mooring in heteronormative spatial matrixes. 'What might it mean to learn to live in the anxiety of that challenge, to feel the surety of one's epistemological and ontological anchor go, but to be willing, in the name of the human, to allow the human to become something other than what it is traditionally assumed to be?' (Butler 2004, 35). To bestow a humane recognition when the gender identity of another is in question is not a simple exercise in tolerance. Nor is it about a strategically timed apology. Humane recognitions are about what Kaja Silverman calls identifications at a distance (1996, 37). Silverman is referring to a mode of recognition that is not based on incorporation (you are just like me), or repudiation (you are nothing like me). Identifications at a distance demand that we recognize difference without aggressive assimilation, abjection, or projective identification. The recognition is structured not by narcissist investments in gender stasis (its purity and legibility in public mirror circuits) but by a willingness to let go of what we take to be true or essential in the domain of gender. Butler notes that the ethical or 'nonviolent response lives with its unknowingness about the Other in the face of the Other, since sustaining the bond that the question opens is finally more valuable than knowing in advance what holds us in common, as if we already have all the resources we need to know what defines the human' (2004, 35).

Sadly, ethical recognitions are routinely withheld in the bathroom. Violence (as I will discuss in a later section) sometimes ensues, and apologies, as suggested above, are often less than straightforward. Sometimes apologies are passive-aggressive – disguising anger beneath a veneer of civility. As Butch Coriander notes,

> I have always perceived people's reactions to be one of embarrassment, so they feel embarrassed that they got it wrong and then they get angry about

feeling embarrassed and then they blame me. People like to be sure of [gender] ... I challenge stuff that they take for granted so they get angry ... I think challenging gender is challenging one of the most foundational aspects of our lives.

Trans interviewees note that they are sometimes read as liars, deliberately deceptive and dishonest about their gender. As Isaac, a white, Jewish, trans man, who is taking hormones (currently in transition), queer, and living in Toronto, explains, 'People feel like ... [those of us] who are gender variant or trans are trying to *trick them*. And the people I know who pass completely as ... male or female can use the opposite bathroom with no problem, because nobody thinks that they're trying to trick them.' Shane, who is a white, post-operative trans male, straight, and living in the San Francisco/Bay area, also refers to allegations of gender fraud and trickery: 'People could be angry if they find out [about your gender], people feel defrauded, and that you've tricked them somehow. And their sense of reality ... is something different, and a lot of times trans people are perceived as being liars, and being false.' Sasha, of Norwegian ancestry, living in Chapel Hill, North Carolina, genderqueer, and bisexual, explains how presumptions about 'gender trickery' are deployed to sanction harassment in toilets:

> It's a gendered space, and so in being in that space, people have implicitly agreed that they are one thing or another. When others start believing that somebody is in the wrong bathroom and that they've lied, are trying to pretend to be someone who they're not, or are in a place that they shouldn't be, then it opens up the door to harassment.[9]

Policing Gender in the Bathroom

> People like having distinctions between men and women. And having those enforced. (Trans man who chooses to remain nameless)

Trans folk often talk about the problem of harassment by security guards and the threat of police arrest. Some speak about assault by non-trans male vigilantes who think they are protecting cissexual women from 'male sexual predators' *posing* as 'women' in the lavatory. Sometimes those presumed to be 'posing' *as* women are, in fact, trans women and sometimes masculine or genderqueer women. These experiences are daily occurrences for some trans folk, whereas for others they are relatively infrequent or non-existent. For some participants, the threat of physical assault and harassment by security guards or arrest by police was described as a relatively constant worry.

Security guards and patrons in general often assume the right to police gender in toilets despite the absence of laws governing the gender of washroom patrons in many American states and in Canada. For example, in Ontario, where the vast majority of interviewees in this study reside, people are protected under the Ontario Human Rights Code. It is illegal to discriminate against, or to harass, trans people by denying them access to public facilities, including public washrooms and fitness change rooms, on the basis of gender. Take the following incident, recounted by Sarah: 'I was at the mall with my sister [in Brampton, Ontario]. And she's, like, "just come to the bathroom [with me]." [I went in and am] ... just kind of, like, standing there, waiting for her, and someone called security.' KJ, who is African-Caribbean and living in Toronto, a trans man (with chest surgery), and queer, tells the following story of a transgender woman and her sister:

> One transgender woman went to the movies with her sister, she was using the female bathroom. And someone, a woman, inside the bathroom saw her and said, 'This is a man. Call security.' Security came in and they harassed her. And she told them, 'I am transgender, you know, this is the right bathroom [for me],' and they [security] made a big scene.

Callum, who is a white trans man, queer, and attending the University of Toronto (U of T), says that his 'ex [lover] was nearly arrested by campus police at U of T in a washroom. Because a woman did a freak out and she ran out ... and it just so happened that there was a campus cop walking by, and he of course came in and there was a big kerfuffle.' Phoebe provides an overview of the problem:

> There is no law in Toronto that says you have to use a specific washroom ... [but] security don't know the laws they're enforcing ... [they] make mistake[s]. At Union station there's been a few incidents ... Someone I know filed a complaint, and had to find legal counsel ... [she doesn't] pass ... her appearance is not working very well for her ... [Her case] ... went to the Human Rights Tribunal, or Commission, and there was mediation, and it was solved ... She received a settlement.

While harassment by security guards was more prevalent than interference by police, the latter was described as more invasive, humiliating, and traumatic – particularly when spectators were present. Consider the following story in which a female-to-male transsexual was arrested by New York police:

> I had just started transitioning and I met this guy who was trans and ... He went to the men's washroom and this cop followed him in and said, 'Let me see your ID—you're in the wrong washroom.' He said that he wasn't in the wrong washroom and that he was just going to use the washroom and that was it ... and then the cop slammed him against the wall, handcuffed him, and dragged him out and arrested him. He had already had chest surgery but they brought him to the women's jail and ordered him to take off his clothes and said, 'Why don't you have any tits,' and all these things to him. So he was very traumatized ... because [the arrest] ... was in Times Square so it was very public. Lots of people saw him getting arrested and being dragged out of the washroom. He got charged with, 'impersonating' and 'trespassing.'

A trans two-spirited person who is Cree and living in Vancouver was wrongfully arrested by police while using the 'women's' room in northern Manitoba. The police were called to break up a fight in the washroom (which did not involve the transgender occupant), and while the women who had perpetrated the assault were not arrested the trans patron was taken into custody. As the interviewee explains,

> I go into the bathroom ... While I'm in the stall this huge fight breaks out ... Based on what I heard ... a group of women purposefully went after a woman that was in a bathroom [stall] ... I heard the police being called ... And then ... I heard male voices. I thought, okay, security is here ... So then I feel it's safe to come out of the bathroom ... I come out of the bathroom only to be grabbed by one of the security guards, and then I get tossed against a wall ... And then the cops come. I then get grabbed again, and I get tossed in the paddy wagon ... I get taken to the police station ... Out of all the women that were in the bathroom, I was the only one that got thrown in the paddy wagon [and then the 'drunk tank'].

While the two-spirited interviewee was not scuffed up or marked in a way that would indicate he had been in the fight, the police proceeded with the arrest.

One non-operative trans woman, Pakistani and living in Toronto, queer, and a sex worker (who has struggled to attain affordable housing), is habitually harassed by security guards in malls when using the 'women's' toilet. As Sarah notes about police, security personnel, and bathroom patrons who have little knowledge about trans folk, 'stereotyping prevails over common sense.' Rico, who is WASP, a non-trans male, and gay, comments on transphobia and gender policing in bathrooms: 'I think there should be more focus on accessibility and less about surveillance.' Rocky, who is a visible cultural minority, genderqueer, and a non-trans female, queer, employed as a server, and living in Buffalo, New York, recalls being spatially segregated from cissexual women in a music-hall washroom by security:

> I was at an Ani DiFranco show ... people were guarding the bathrooms ... [security] guards stopped the line ... they decided that they needed to make sure that people were specifically segregated [by gender] and the line was stopped when I went into the stall. I went to the women's room because I figured it wouldn't be a big deal ... [but] the line was stopped and held up until I was actually out of the bathroom. I was infuriated ... it was embarrassing. Nothing was said, there was no verbal ... communication ... But spatial politics came into play.

Non-trans men other than security guards sometimes interrogate and assault trans women and masculine-presenting people under the auspices of chivalry and protectionism. Consider an incident narrated by Bryan, who is of Irish descent, white, transmasculine, queer, and, at the time of writing, coordinator of the Safe Bathroom Access Campaign for the Transgender Law Center in San Francisco:

> I have a friend who is a trans woman who was using the ['women's] restroom at ... a gas station ... there was another [cissexual] woman in the restroom with her ... when my friend came out of the restroom ... [the other woman's] boyfriend was waiting for her at the door and punched her ... [he] ... was upset by her being in the bathroom ... many times, men are policing women's rooms.

As KJ notes, 'It upsets me that people are so vehemently attached to the fact that ... [bathrooms] have to be segregated ... it's almost like a policing of the public washrooms ... [by] ... vigilantes [and usually men] ... No one's asking them to do this.'[10]

Rohan relates a personal experience of assault:

> I got accosted by some straight guys coming out of the women's bathroom one time and subsequently got my ass kicked because they ... probably thought I was a dude ... You have to worry about people seeing you coming out and identifying you as queer.

Gypsey speaks about her human-rights lawsuit against a Vancouver bar owner. Although she explained to the male bar owner that she lives as a woman and identifies as transgenderist, she was still denied access to the 'women's' room in the bar. 'The owner said: "You are a man because you've got muscles, you've got this, you've got that, and you've got this." And then he said: "How is your plumbing, what kind of plumbing do you have?" '

Gender and Safety in the Water Closet

> I think gendered bathrooms are inherently unsafe. (Interviewee who chooses to remain nameless)

The gendering of public facilities is often rationalized by dominant cissexist and heterosexist safety narratives. Interviewees say that non-trans people invested in heteronormativity want bodies sorted into oppositional categories – male and female – allegedly for physical safety and security. Dominant safety narratives (which are challenged as they are also used and reworked by interviewees) maintain that women have legitimate fears of assault, that assailants are usually men, and that gender-exclusionary space affords more protection than gender-inclusive spatial designs. Even the most feminist and trans-positive safety narratives play a role in constructing the very violent acts (and people) they describe. As David Valentine (2007) argues, narratives about violence are less than transparent. This is true when the accounts are relayed by cissexuals *and* by trans people. While there are very real accounts of physical violence in the interview data (which I believe should be documented), there is also a way in which the telling of a violent story is shaped by narrative structures dependent upon a 'category of social identification, a category which in turn requires the complexity of violence and its multiple structural logics to be smoothed over' (228–9). In the telling of violent stories, I want to do justice to the interviewees. But I also want to think critically (as many interviewees do) about what counts as violence, about who can and cannot be violent and subject to violence. It is crucially important not to reproduce hegemonic safety narratives that obscure forms of violence that do not fit the dominant cissexual template. There is an antiquated and heterosexist construction of masculinity underpinning cissexual safety narratives. For instance, one transgender interviewee observes that 'there's a whole idea in this society that ... if a man sees a woman, just a glimpse, he cannot be controlled.' Men are constructed as impulsive and predatory, whereas women are constructed as potential victims. One trans man observes what he refers to as 'a continuing feeling that women need to be protected from the sexual advances of men; if they shared the same bathrooms, men would always be peeping in on the women ... that's the general perception.' Chloe, a white, non-trans queer femme living in Vancouver, contends that 'It's so ingrained into women that men can be potentially violent. And that one should not feel safe around a strange man.'

While trans and cissexual women alike have good reason to worry about personal safety, the dominant narrative construction of danger is gendered in ways that reproduce these very dangers. Cissexual and transsexual women often say they are reluctant to share bathrooms with non-trans men. But those interviewed also realize that isolated and compartmentalized spatial designs that segregate people by gender set up the conditions for – and do not guard against – physical and sexual assault. Emily critically reflects upon her worry about gender-inclusive restroom facilities and identifies what she takes to be illogical in her worry:

> I am not completely sure that I want to be alone in a bathroom with a man ... I have a fear about that. I think that's my main objection to unisex bathrooms ... On the other hand ... it's [more of] a perceived danger than a real danger because ... there isn't any magic formula that stops the rapist from going into the women's bathroom ... We tend to think of it as magic armour that you can't get through but really there isn't anything stopping them.

Rohan, who is trans and butch, makes a similar observation:

> The assumption is that if you are in a women's bathroom and you are a woman, then you are 'safe' from men who rape, or stalk or both. Of course it's an illusion, because men who rape aren't going to respect the sign on the door ... In some ways a washroom is an ideal place to do that, because it's out of the public eye. It's in public but it's out of the public eye.

Gendered and enclosed architectural designs are, as Rohan conjectures, ineffectual when it comes to violence prevention.[11]

A second problem with heterosexist and cissexual safety narratives is that some people are imagined to be aggressive and dangerous while others are imagined to be potential victims. This presumption obscures patterns of violence that do not fit into dominant danger narratives. It constitutes masculine people in the 'women's' room, along with recognizably trans woman in that same room, as always already potentially violent. Too often, trans men and trans women, butches, genderqueers, and those who are gay, lesbian, or bisexual are believed to be predatory, while violence against LGBTI people is rendered invisible and goes unreported. One transgender interviewee notices that 'people cannot always distinguish between a male sexual predator and a masculine woman or transgender man.' As a femme interviewee (whose trans or non-trans status is unspecified) says, 'You know someone who is not clearly feminine is reduced to the man perpetrating some kind of perverted activity in the bathroom ... they are aligned together. It's not justified but that's just the way it is.' Jacq also notes a widespread confusion between a male sexual predator and those who are butch in the 'ladies' restroom: 'When they [women] see some big butch woman walking in ... [there is panic] in women's faces. [There is] sheer *fear* when I walk in the door ... They're afraid that I'm going to hit them or be violent in some way or say something.'

The confusion about who poses a physical threat and who is using the 'women's' bathroom for legitimate purposes is questioned by some trans men. As one trans man notes, 'I am five foot one ... who am I going to hurt?' Property, who is Caucasian, living in Toronto, and identifies as a queer female who has discontinued taking hormones (which she originally took because she intended to transition from female to male), wonders why fears of masculine people are not offset by large numbers of women in the bathroom: 'The other thing, too, that I never understood – when women were screaming about me being in the bathroom, was [the following]: there are TWELVE of you in high heels. If I did anything right now, I'm so going to get my ass kicked! I'm going to be dead!' Callum also reflects upon the peculiar composition of the fear-based response had by cissexual women:

> It was like they felt unsafe ... They thought there was a man in the women's washroom, that I was dangerous. Which is kind of bizarre because [the room is] ... *filled* with women ... Maybe it's more about this is strange, and not about a personal threat. So I guess it's not just a safety thing.

As Rohan speculates, 'I think that that need for [a safe gendered] space for some people gets twisted into a really nasty sense of entitlement, to interrogate me, to interrogate other trans folks, other butch dykes.' Farah, who is Caribbean-Canadian (originally from Trinidad and Tobago), transgender, queer, and living in Toronto, believes that what *seems* to be a fear-based response (driven by a worry about sexual assault) is actually a transphobic or homophobic response:

> I've heard stories of transgender women ... not passing [and being seen by non-trans women in the toilet]. And instead of ... [leaving because they are] scared ... what they've done is gone up to the transgender person and screamed in their face. Now tell me ... if you're afraid of someone would you go up and scream in their face, 'What are you doing here'? Would[n't] you run [and] ... try to escape? So is it really about fear or is it transphobia or homophobia? So are [cissexual women] really afraid for their safety? If I [were] ... thinking that someone's going to attack me, I'm ... [going] to go in the opposite direction. I'm not going to get up in their face!

As suggested by interviewees, the pretence of fear sometimes masks gender panic and aggressive disidentifications. The same allegations authorize verbal aggression towards trans people. In essence, there is a slippage between what may, in some cases, be a legitimate fear of assault and an aggressive transphobic response. Worries about sexual assault are sometimes fuelled by gender panic. One's own felt gender identification and investment in sexual difference are at risk. Transphobic angst is clearly encapsulated in the following account about a transgender woman friend of the interviewee:

> She's recognizable as a trans woman, she definitely does not look like a man. Long hair, very feminine, dresses, skirts, everything ... And ... [a patron looked at my friend] ... and looked at the sign on the door, and then looked at her again ... [People like this patron] ... would say, 'Oh I was afraid because I thought it was a man.' But it's actually, like, no, you thought it was a trans woman. She didn't look like a man. Men don't wear ... long skirts and wear make-up. This girl dresses in high heels. She looks like a trans woman, right? And that's where it's not about danger. It's about, like, who is normal.

Gender panics are ignited when gender identities are perceived to be in discord with the sex of the body. Those who are seen to defy the normative cissexual gender order can be subject to aggression and hate. As one interviewee notes,

> Women [do] fear ... assault ... Which is a really valid fear, but at the same time, it doesn't always look like fear. Sometimes it actually looks a lot more like hatred. Where women would go up to someone, [a] trans woman, trans man, or butch ... and scream in their face, 'You're disgusting. You don't belong here.' And that's not actually what you do when you're afraid.

Interviewees also explain how these expressions of hate are gendered. As one transgender interviewee notes, 'If I go into a women's washroom, the worst thing that happens to me is, I freak someone out and ... this awkward moment [follows] ... But if I go into a men's washroom and I don't pass, like, I can get beat up or raped or something.' Isaac says that he would rather use the 'women's' room because it is safer but

> started using men's bathrooms ... because [of] ... the interrogation [he received post-transition] ... It came to the point where ... *every single time* I went into a women's washroom, someone would say, 'Are you sure you're in the right bathroom?' Or, 'You're in the wrong bathroom!' or they'll turn to someone and say, 'There's a man in here!' ... I get much less looks ... in the men's bathroom than I did in the women's, because men don't look ... they don't look because that's what fags do ... gay men look.

Interviewees suggest that transphobic violence in the 'men's' bathroom is triggered by non–trans-specific heteromasculine fears of feminization and sodomy. Jacq, who uses both toilets, notes how people who are read as 'feminine' in the 'men's' room risk assault because they are presumed to be 'gay': 'There was ... one guy that got beat up because he looked a bit femme. He may not have been gay. I know people who are not gay who look feminine ... [who get] hassled ... using the bathroom.' Sarah comments on experiences of abuse in the 'men's' lavatory, and how the assaults are precipitated by how her sexuality is read as 'gay': 'I've ... been ... spit at, had something thrown at me, been assaulted ... [subjected to] hate-crimes ... ninety-five per cent of the time, it's men. And ninety per cent of that ninety-five, they think that I'm a gay man.'

We may understand violence in the 'men's' room as symptomatic of gender and sexual panic. A non-trans man who physically assaults a trans guy, a cissexual gay, or an effeminate man may do so because he feels his gender identity and/or sexuality to be at risk. The violent response is often an attempt to police the cissexist and heteronormative composition of the room. It may also be a desperate attempt to project one's own identificatory injuries and anxieties onto another who is more easily aligned with social abjection (or, rather, treated like shit). Judith Butler suggests that the 'negation, through violence, of that body is a vain and violent effort to restore order, to renew the social world on the basis of intelligible gender, and to refuse the challenge to rethink that world as something other than natural or necessary' (2004, 34). To sidestep the violent response is to live with ambiguity, to increase our tolerance for risk along with that which we do not recognize in the mirror image of others. 'The violent response is the one that does not ask, and does not seek to know. It wants to shore up what it knows, to expunge what threatens it with non-knowing, what forces it to reconsider the presuppositions of its world, their contingency, their malleability' (ibid., 35). A willingness to live with our own incoherence is vital if we are to affirm a space for the other who may not return a gender-normative image. The non-violent response is one that can afford to abandon a heteronormative and cissexist quest for gender purity. Trans-positive ways of seeing require us to expand and reconfigure gender norms so that they are less exacting and restrictive to those who can access cissexual privilege. In chapter 3, I consider how visual economies of power operate through the panoptic and mirrorical designs of the lavatory, surveying gaps between the body, its genitalia, and gender presentation.

3 Seeing Gender: Panopticism and the Mirrorical Return[1]

> In assault cases that I've heard of ... the head bashes against the mirror ... Or [it] get[s] ... cut with the mirror.
> (Erik, who is Ojibway and transgender)

> What makes a public bathroom safe or unsafe? I think how the user of the bathroom is *perceived*.
> (Gypsey, who is transgenderist and lesbian)

Michel Foucault (1979) argues that the modern era inaugurated a new form of discipline: the decline of corporal punishment as public spectacle and the rise of surveillance. 'The classical age discovered the body as object and target of power' (136); it became an object of visual scrutiny. Elimination also became an area of discipline, but the sanitary technologies designed to survey the body and its toiletry habits were not perfected until the late nineteenth century. Sir Edwin Chadwick, the English sanitary reformer, argued in his now famous report *The Sanitary Conditions of the Labouring Classes in Great Britain* (1842) that the unclean and the immoral had to be staged in plain view. Their toiletry practices had to be made visible to promote public hygiene and to police elimination – the indiscriminate depositing of excreta onto city streets and into illegal cesspools. The architectural divide between the lower and upper classes was orchestrated, in part, through hygienic appeals to sanitation reform and to a concurrent wish to subject the former to visual inspection and governmental regulation by the latter. Modern architectural designs – from the prison to the now public lavatory – gave new and unprecedented attention to what Foucault calls the 'anatomy of detail' (139).

From the Middle Ages through to the early modern period, elimination was less organized and more communal than it came to be in the late-Victorian era. Promiscuous urination and indiscriminate defecation were commonplace. As Kitchen and Law (2001) remind us, 'Prior to the period of [the] Enlightenment in Europe, urinating and defecating was [sic] a public act, taking place in fields and gardens, but also in the street. By the mid-19th century, however, it had become a private act, taking place in out-houses or inside the house, and was disciplined and socially regulated by Victorian reformers championing modern sewage systems and their promotion of discourses of public health' (288). Toiletry habits could be disciplined by panoptic design. Public toilets in England and many European countries – formerly thought to be vulgar and indecent – came to symbolize the 'clean' city; they were markers of national pride and measures of civilization.

Each water closet authorized a distinct choreography – what to do and how; into what receptacle, oval pedestal or urinal; and in what order. The choreography of the body – how it shat and pissed – was a testimonial to one's class and genital organization. All-seeing eyes narrowed in on gaps, the distance any given body took from a normative excremental ideal. By exposing a gap, a perceived difference, between any given body and the norms deployed to govern it in the lavatory, people were subject to a new visual economy of power focused upon the body, its genitals, and its gender as measured through toiletry habitus (Inglis 2002).

Gender norms in the present-day public lavatory are enforced by panoptic designs dating back to the early modern period. As Foucault explains, a panopticon is a modern technology of power based upon the concept of the prison watch tower, an architectural design drawn by Jeremy Bentham in the nineteenth century. 'The Panopticon is a machine for dissociating the see/being seen dyad: in the peripheric ring, one is totally seen, without ever seeing; in the central tower, one sees everything without ever being seen' (Foucault 1979, 201–2). While there is no absolute concordance between the prison watch tower perfected by Bentham and the many disciplinary institutions built in the modern era (including the public lavatory), there is, as Foucault contends, a common use of optics. There is what he calls a diffuse 'web of panoptic techniques' (ibid., 224) that involve normalizing judgment. 'Thanks to the techniques of surveillance, the "physics" of power, the hold over the body, operate according to the laws of optics and mechanics, according to a whole play of spaces, lines, screens, beams, degrees and without recourse, in principle at least, to excess, force or violence' (ibid., 177).

Ways of looking without being seen are central to panopticism. The look emerges from a particular point, but it cannot be intercepted. One is never sure if he or she *is* seen at any given moment, although one knows full well that he or she *can* be seen. Panoptic designs in Canadian and American public restroom facilities of today are created by the assemblages of right angles, smooth lines, refractive tiles, glass, and other mirror-like materials, and by the metallic compartments and the vertical positioning of stall partitions and doors. The rooms are voyeuristic spaces functioning to incite wonder and intrigue while maintaining a pretence (however dubious) of privacy. Power over the body and its eliminatory rituals is enabled by architectural designs cultivating scoptic pleasures in the detection of anomalous positions and genital configurations.

Stallybrass and White (2007) note, in the case of London, England, that the 'vertical axis of the body's top and bottom is transcoded through the vertical axis of the city and the sewer and through the horizontal axis of the suburb and the slum of East End and West End' (282). As the lower strata of the bourgeois body become hidden or 'unmentionable' in the Victorian era, we 'hear an ever increasing garrulity about the *city's* low – the slum, the ragpicker, the prostitute, the sewer – the "dirt" which is "down there" ' (281). The horizontal is aligned with the feminine, the city's underground, crime, vice, death, disease, and degeneracy. The horizontal is subordinated to the vertical and to the visible. The upright position is metonymically associated with the masculine, the 'good,' the 'clean,' the 'healthy,' the 'above-board,' and the 'upstanding' citizen.

Lavatory designs in North America typically prompt men to stand and women to sit. This is one of the ways gender is visually apprehended. The vertical body is erect and discernible, masculine and autonomous. The horizontal body is feminine and relational, unstable, leaky, or ill-defined in the hygienic (and phallocentric) imagination. By architectural design the feminine is put in close proximity to base body matter (or city sludge); she (or he) is required to sit and to make direct contact with the oval receptacle. The urinal and the oval toilet bowl authorize dimorphic urinary positions – standing or sitting – in accordance with cissexist presumptions about gender and genitals (even when many people may prefer to squat).

The gendering of the bathroom is to be found not only in its receptacle designs and outer-door insignias but in the disdain for abject body matter coded as feminine built into its design. A good example of the disdain for bodily remains (coded as feminine) is to be found in the recommendation made by Alexander Kira (1966) that in order to promote hygiene and sanitation, architects and builders in New York City should focus on the construction of surfaces that can be easily cleaned and upon materials that are non-absorbent. He advocated 'structural soundness, dimensional stability, chemical stability and inertness, abrasion resistance, stain resistance, non-absorption, freedom from odor retention, visual and bactericidal cleanability … [and noted that] it would be desirable if the colors and patterns were such as to show up dirt' (89). He favoured vertical lines, light colours (preferably white) to make dirt visible, well-ventilated closets (with short doors that did not reach the floor), and noise-proof construction (so sound could not rebound and echo). As noted by Lahiji and Friedman (1997), 'The vertical of the right angle dominates not only the tradition of architecture but also the ethics of modernity' (46), which is, according to Luce Irigaray, also masculine. She contends that modern architecture reveals an intolerance for the feminine through its disdain for maternal (or intimate) public space, diagonal lines, curvature, fleshy or ill-defined surfaces, fluid passageways, plurality, sense-based experience (other than visual), holes and receptacles, asymmetry, infinity, and excess (Rawes 2007). In short, modern architectural design is 'unsexed' (to use Irigaray's word) in its disavowal of the feminine (and, I would add, of visibly trans and gender nonconforming folk), and in its wish to refract homogeneity or, in the case of the toilet, a gender-neutral sameness – each room is, after all, strikingly similar, despite superficial differences in receptacle size, shape, and design.

Gender stasis is architecturally mandated in the present-day Canadian and American public lavatory. While there is no one bathroom warden, no single tower or centre of surveillance, there are multiple eyes in public lavatories (nowhere and everywhere, concurrently). Gender is subject to surveillance in ways that depend upon light, optics, and fixed choreographies (as in the Bentham prison), but the patterns of surveillance in the lavatory are unique insofar as they are refractive, triangular, and uncanny. Gender recognitions in the modern toilet are negotiated by what Helen Molesworth (1997) calls the 'mirrorical return' (81) in her discussion of Marcel Duchamp's *Fountain* (a controversial installation of an inverted urinal entered into the American Society of Independent Artists exhibition in 1917). While the 'mirrorical return' originally referred to the splash of urine in the inverted urinal, it is now used as an analytic concept in academic theories of the modern subject. If the masculine subject is dependent upon the urinal to contain and wash away body waste, the inversion (fountain-like shape) of the receptacle undermines the unity and order of the virile and autonomous masculine body. To be reunited with abject body fluids is to be reminded that one is not whole but haunted by part objects. 'Here the concept of the "mirrorical return" problematizes the wholeness of the producing body, for body and fragmentary waste product would be rejoined, complicating the whole or part status of either' (ibid., 82).

The mirrorical return may also designate a psychic process through which we negotiate the return of body parts as images. Identifications are forged upon the exchange of attributes we do and do not want. Kaja Silverman defines the 'self-same body' as a psychic structure through which 'the ego consolidates itself by assimilating the corporeal coordinates of the other to its own – by devouring bodily otherness' (1996, 24) in a narcissistic fashion. Conversely, we reject body parts and images that we cannot or do not want to assimilate because they are at odds with our psychically invested corporeal coordinates and internal self-imagery. The repudiation may involve an all-out negation of the other in the service of a coherent gendered ego ideal. The repudiation may also lead to the construction of the de-idealized other as abject through a hyper attention to the difference signified by the other. What Silverman calls 'alien corporealities' (ibid., 24) are refused by the one who feels proprietary about gender normativity and the identifications it engenders. The mirrorical return is a dynamic of exchange in which gender ideality or normativity is in dispute and up for grabs. Diana Fuss (1995) reveals that Freud's notes on identification involve a

> spatial metaphorization of the unconscious as a field of divisions, hostilities, rivalries, clashes, and conflicts ... [And that in the spatial field] objects that cannot be kept out are often introjected, and those objects that have been introjected are frequently expelled – all by means of the mechanism of identification. (49)

The spatial field becomes a borderland where part objects are assimilated or expelled. The lavatory is a contested and angst-ridden cultural terrain because it reminds the subject of what has been forgotten (or refused) in the making of the gendered self. It is not a site that enables us to see the 'otherness of the desired self, and the familiarity of the despised other' (Silverman 1996, 170). The toilet is a fractious terrain in which the ego patrols the gap, the distance or fissure between any one body (or part object) and the self, between a normative gender ideal and the other as a mirrorical object.

Jacques Lacan (2006) tells us that 'all sorts of things in the world behave like mirrors' (Seminar II). People in the bathroom behave like mirrors; architectural shapes and material objects, including reflective glass and metallic surfaces, all act as mirrors. Space is not neutral or passive. According to Lacan, space looks back at us, and we are constituted by the way we are apprehended or caught in space. In his psychoanalysis of space, Henri Lefebvre (1998) explains that 'One truly gets the impression that every shape in space, every spatial plane, constitutes a mirror and produces a mirage effect; that within each body the rest of the world is reflected, and referred back to, in an ever-renewed to-and-fro of reciprocal reflection' (183). By using the glass mirror alone, one can see, and be seen by others, in a reciprocal and triangular fashion. Our bodies may be apprehended by a single, wide-sweeping glance in a vertical or elongated looking-glass. We see how we are being seen, and the triangulation of the panoptic eye is multiplied by the number and kind of mirrors in the room.

We are no longer talking about a single eye or point of surveillance as in the Bentham prison proper, but of a splintered and triangular panopticon. In visual economies of power, glass mirrors are centres or vestibules of surveillance, but the play of power is refractive, diffuse, and impossible to pinpoint. If identifications are, as Fuss (1995) suggests, a 'compromise formation or a type of crisis management' (49), things that can be 'dislodged by the newest object attachment' (49) and re-entrenched by repeated disidentifications and projections (deflecting unwanted attributes of the self onto others), then it should not be surprising that the room (curiously privatized in public space) where we shed fluids and open orifices should also be the room where gender identities are felt to be especially fragile. Panoptic designs in the toilet reveal gaps, or rather a lack of concordance, between gender identity and the sex of the body apprehended by mirrorical eyes. Transphobic attention is focused on the perceived discord.

Gender under the Microscope

Trans interviewees unanimously agree that they are thrust into a paradox whereby they are either subject to visual scrutiny in bathrooms or rendered invisible. In each case, the glass mirror plays an important role in determining how one's gender is intercepted. Many participants talk about the colouring optical design of toilets. Interviewees feel that by making gender ultra-visible, bathrooms put bodies on display.

Diane, who is a white, non-trans woman, androgynous, lesbian, and employed as a lawyer, says that bathrooms 'tend to be a very industrial ... beige and grey ... [It's] a sterile cleanliness ... subconsciously ... there is something about ... the colour white ... like a doctor's lab coat ... [it is] hygienic or ... sterile.' Gender examinations are enabled by optical designs of the toilet inviting surveillance. The eye scans for anomalous bodies, permutations of gender at odds with the binary signage on the washroom door. Interviewees compare the bathroom to a 'microscope,' to a 'photo studio,' and to a 'scientific lab.' Rohan, who is transmasculine and butch, says, 'I feel like I am under a microscope every time I am in a bathroom.' Velvet Steel, who is a post-operative trans woman, offers a detailed sketch of what Foucault would call the panoptic technique of power established through disciplinary partitioning:

> Once ... you get ... to the sink ... that's when people ... really have an opportunity to scrutinize you and look at you because ... we're standing there with these bright lights in front of a huge mirror that usually takes up the whole wall. That's projecting your image *back* to the rest of ... [the people in the room] ... People can stand there and look in the mirror and look around the rest of the washroom, you know, at the cubicles, the stalls, the toilets, whatever ... [you can] have a *gander* at what else is going on in this place ... You almost feel like you're on display! ... Your regular run-of-the-mill, generic public washrooms are ... microscopes in a lab ... very sterile ... there's a logic ... behind the design.

As interviewees suggest, there is, built into disciplinary institutions theorized by Foucault, a 'machinery of control that functions like a microscope of conduct' (1979, 173). These disciplinary architectures were heralded in by new technologies such as the 'telescope, the lens and the light beam' (171). What Foucault calls the 'art of light' (171) is central to modern optical designs. As Madison, who is an Italian-American living in New Jersey and genderqueer, notes,

> I think that the colour of them [bathrooms] lends itself to an institutional-type feel ... it's often this fluorescent type lighting that when reflected off white surfaces ... [like] sinks [which] are white ceramic ... [and] the [wall] tiles are, like, white or beige, the doors themselves ... [are] 'neutral' colours ... I think that the fluorescent light really adds to that institutional ... deer-in-the-headlights, Seven-Eleven at three in the morning-type lighting.

People are unnerved, or caught like a 'deer in the headlights' by a bright, fluorescent 'light' placing the body in a spotlight.

When asked about the architecture and design of public restrooms, most interviewees comment upon the bright white lighting, the sanitary texture or 'feel' of the room, and the 'cold,' 'hard,' and 'impersonal' metallic surfaces. More than a few reflect upon the racialization of the public toilet as white. For example, Emily, who is intersex and lesbian, notices that bathrooms are designed to 'foster the illusion of sanitaryness, by things being white. I'm not sure we always think through the racial implications of that.' Tara, who is white, genderqueer, and gay, believes that 'white has a psychological ... association with cleanliness.' Neil, who is a trans man and a person of colour, challenges the presumption that whiteness should be indicative of cleanliness, thereby underscoring the racism driving the conceptual equation. He says that the pairing of 'black' with 'dirt' is 'just awful ... if you are thinking about ... [it] in terms of race.'

If the history of the white, bourgeois body that Foucault tells can also be a history of empire, colonization, and modern racisms, as Ann Laura Stoler (1995) contends in her reworking of Foucault's history of sexuality, then it is essential to read disciplinary institutions as playing a pivotal role in the creation of 'race.' 'One could argue that the history of Western sexuality must be located in the production of historical Others, in the broader force field of empire where technologies of sex, self, and power were defined as "European" and "Western," as they were refracted and remade' (195). A truth about sex or gender is dependent upon techniques of racialization producing interior and exterior others. One of the key points made by Stoler is that modern bio-politics enables one to 'distinguish the interior Other and know our true selves' (190). Gender and race are both consolidated in public mirrors; internal and external others are subject to visual surveillance.

Whiteness – as a colour and a racialized aesthetic denoting purity, hygiene, and public sanitization – enables the glass in the hanging mirror to refract. White backdrops function to stage gender. For trans interviewees, glass and public mirrors highlight a lack of congruence between the body as it is felt and the body as it is intercepted by others. Cognizant of how their bodies are subject to surveillance, most trans people are ambivalent about the mirror hanging over the sink. As Jay Prosser (1998a) suggests, 'for the transsexual the mirror initially reflects not-me: it distorts who I know myself to be' (100). The sphere highlights the gap between the internal imagining of one's body (the visual imago), and the externalizing projection of the body as seen by others.

Trans interviewees note the distortion felt before the looking-glass and refer to it as a 'beacon,' a 'centre of surveillance,' and even compare it to a prison 'watch tower.' Many are anxious about being 'caught in the fluorescent spotlight.' Others are unsettled by how their images are projected into space from multiple angles. As one interviewee (who chooses not to provide demographic information) explains,

> Mirrors ... allow ... people to sort of watch you from different angles without actually turning around and looking at you. So ... if you are already in a space that you know you're otherized in, then ... the existence of a mirror becomes even more of a way to police and make sure your body looks just the way it's supposed to look.

One queer and non-trans graduate student believes that 'mirrors are the site where the panopticon is actually operating.' Another queer femme who is not trans confirms this view: 'mirrors are definitely used for ... surreptitious [gender] surveillance.'

The modern lavatory is an institution where visual economies of power are operable. But where there is power there is resistance (Foucault 1978), and looks in the toilet are subject to mirrorical returns. Trans people often interrupt and refuse inhumane modes of address that draw unwanted attention to a gulf between gender identity and the sex of the body. They interrogate, and exchange what Kaja Silverman (1996) calls a de-idealizing gaze. A gaze that undoes one's claim to humanity along with one's right to gender self-determination is de-idealizing and transphobic. These looks are subject to resistance and return. But this resistance is never exterior to power relations giving rise to injurious modes of address. As Butler explains,

> As a further paradox, then, only by occupying – being occupied by – that injurious term can I resist and oppose it, recasting the power that constitutes me as the power I oppose ... any mobilization against subjection will take subjection as its resource ... [as] an attachment to an injurious interpolation. (Butler 1997a, 104)

There is resistance to the disciplinary machinery, but this resistance is not outside the field of power; it works within it, redeploying its terms and neutralizing its capacity to injure.

A good example of resistance and redeployment can be found in the use of the mirror in the 'women's' room. The hanging glass is often used for self-protection by the very people subject to disparaging looks.

Velvet Steel, who is a trans woman, elaborates on how she uses the mirror as a personal security device: 'I'm using the mirrors in the washroom more as a security measure for me to see what's going on ... I will use whatever tools [are] available to me, such as the mirrors [because] ... I don't have eyes in the back of my head ... [I try to avoid getting into] the reflection of the ... mirror ... [this provides] a good sense of ... security for me.' Interviewees also talk about 'catching' or 'intercepting' people who give transphobic looks in the mirror. When de-idealizing looks are returned by the one who is meant to be shamed and effaced by an unflattering gaze, the voyeur is, not infrequently, startled. Interviewees often refuse (or return) disparaging looks. Rohan (who was physically assaulted upon exiting a public restroom) comments on being 'seen' in the mirror:

> People use mirrors to watch me. So they won't actually turn and look at me; some people will. They won't actually ... spin around and stare at me and go 'Oh my god, what are you doing in here?' But they will covertly watch me out of the corner of their eye, and I can feel it ... I will turn and look in the mirror and catch them doing it, and they ... jump and ... [act] like 'Oh you caught me watching you.'

Non-trans people frequently use the mirror to give transphobic looks. Those giving injurious looks do so under the pretence of anonymity and invisibility, as indicated by Rohan. In Madison's words, 'People feel like they can look in the mirror at me. Because they're not looking directly at me, like, they have access to staring that's not actually looking at me. They're looking through the mirror which somehow ... makes it okay.' Callum, a trans guy, also notes, 'Well [mirrors] they're supposed to be to look at yourself and check yourself out. But it's interesting, because people, they're watching you without watching you, right?' As interviewees suggest, in public spaces with a glass that doubles, or reflects, *in the place of the self*, the mirror obfuscates an interrogating visual exchange that might otherwise be more direct and explicit. Chloe, who is a non-trans lesbian femme, says that when she and her transgender partner are given aggressive 'looks,' she has 'taken to staring right back at them. And almost in a very confrontational way with my eyes, because I would like them to know how it feels.' In this mirrorical return there is a queer or trans dynamic set in play by the returns on looks that disturbs cissexist and heteronormative mirror circuits.

Because transphobic hate is not all that it seems and disidentifications are, as Butler (1993) suggests, sometimes the product of unconscious identifications already made (however repressed), it should not be surprising that what is seen to be aggression in the exchange of looks can slip into the terrain of the erotic.² There is a fine line between love and hate, desire and disgust, recognition and its refusal. Transphobic looks such as those cited above are not just hostile and persecutory (although they are), but sometimes soliciting. Interviewees who use the mirror to survey bathroom patrons for self-protection are acutely aware of how easily their 'looks' can be read as provocations, or as invitations to sexual intimacy. Sometimes cissexuals who are transphobic project (or transfer) their own personal troubles with gender (or sexuality) onto trans people. Those who are recognizably trans and/or gender non-conforming (or presumed to be gay, lesbian, or bisexual) are sometimes read as soliciting sex. As Sarah, who is transgenderist, notes,

> If I'm looking at somebody, it's generally ... [because I am] assessing whether or not they are a threat. Like assessing the situation as opposed to ... looking *at* them. I look for little things, like if they're ... fidgeting or twitching or if they're looking at me ... I pay more attention to body language ... [Other than that] I actually try and make a specific point of not looking at women when I'm in the bathroom because I look so gay. I don't want them ... [to think] '*Augh, she's looking at me* [in a sexual way].'

Chloe says that, 'It has crossed my mind a few different times to avoid eye contact with absolutely anybody there. To just be one hundred per cent sure that I do not give off any sort of ... unspoken ... [sexual] signal ... [before] I then proceed into a public place and pull my pants down.' Rohan says, 'Usually, I just look at the ceiling.'

Some trans folk are, by contrast, compelled to look at other people to legitimize their claim to the 'women's' room. Neil says that, 'In the women's washroom, you walk in and you want to make eye contact with every female form on the floor to sort of validate the fact that you belong here. It's very awkward.' As another interviewee who is a trans man explains,

> If you just go in and go into the stall and you don't make *any* sort of eye contact and you don't participate in *any* of the social niceties *they are going to call the police*. And the police are going to come or you are going to get the security guard or whatever the hell – bouncer or whoever ... you have

to normalize the fact that you are there and you *should* be there and if you don't you're in *trouble*.

The same look given to avoid trouble can also, as suggested above, incite trouble if one is (wrongly) presumed to be signalling sexual availability. As trans interviewees note, one is often read as either an imposter, guilty of encroaching upon 'women's' space, or as sexually provocative and predatory.

In the 'men's' toilet different optical strategies are employed to avoid gender-based interrogation. In this room, it is often necessary to refuse eye contact altogether. Occupants have to be quick and task focused. *Thou shall not linger or 'look' at other men.* Delicate and careful hand washing and peering into the mirror for any length of time are discouraged. As one interviewee, who is a trans man, reveals, 'I usually recommend [to trans men who are newly transitioned or transitioning] … Go to the men's bathroom, look at the ground, grunt. That's all it takes.' As another transgender guy from the San Francisco/Bay area explains, I am a 'little bit more brusque when I use men's rooms … I rip the paper towel out of the thing [dispenser] and [make a loud noise] …' The same interviewee from the Bay area notes that 'the most horrifying behaviour in men's rooms would be friendliness.'

Mirrors are frequently missing or broken in the 'men's' toilet, a fact that may indicate an active resistance to a panoptic gaze. It may also indicate a heterosexist refusal to be a homoerotic spectacle for other men. Vanity is also associated with femininity and interiority and seen as contrary to white, normative, cissexual, heteromasculine self-definitions. To be seen peering into the mirror is to risk being read as gay and/or effeminate. A trans man (who doesn't provide demographic information) insists that 'women's bathrooms have more mirrors. Men's are more likely to not have one.' Another interviewee says with exasperation, 'Women have a huge-ass mirror in front of all the sinks and [the] men's [bathroom] tends not to [have a mirror] … I [think that] if women had a urinal then men would have a mirror.' Callum observes the same homosexual prohibition on looking into the glass: 'When I'm in men's washrooms I don't really see a lot of men preening … and when I do, I feel it's a bit gay. I feel a bit suspect for doing that. In, like, a straight sort of identified area. Men are hardly ever looking in the mirrors.'

At the same time, participants note that these masculine gender protocols are mediated by class and venue, and that cissexual men who are heterosexual do sometimes peer into mirrors to adjust a tie or shirt or to comb their hair, particularly in restaurant, bar, and theatre restrooms. It does, however, seem to be the case that mirrors are more likely to be missing, cracked, or torn off the wall because of the comparatively stringent prohibitions against looking in the heteromasculine space.

A lingering gaze at the urinal is, often, said to incite homophobia. Interviewees believe that cissexual hetero men do not want their penises subject to homoerotic looks. It is thus important not to look at all. As Mykel, who is a white, Anglo-Saxon, non-trans gay male, tells us, 'You are aware of what's transpiring around you, certainly [in the men's room], but there wouldn't be direct [eye] contact unless the individual is looking for sex or something like that.' Most trans interviewees agree that to properly navigate the 'men's' room one must not look at other men (or, at least, not get caught peering downward before a urinal). As another interviewee confides, 'Given that I look pretty queer, I know … I would risk getting beat up … if I got caught [looking at another man's penis while at the urinal]. And anybody who saw me do it would assume I was queer.' KJ, who is a trans man, confirms this: 'Definitely [men] don't look at each other's penises … it comes more down to [the threat of] homophobia.' To be read as a cissexual, it is often necessary to abide by prohibitions on homosexuality. Neil explains:

> You go into the men's bathroom, you don't make eye contact. You walk in, do your stuff, and walk out. I've only ever bumped into a guy in the men's bathroom once and even then he was on his way out and barely looked at me … the less eye contact and the less conspicuous you make yourself seem [the better], just stare at the floor purposefully and walk to the place you need to do your stuff, you seem to pass more as male.

A refusal to look is, however, never an absolute refusal. There are numerous testimonials to confirm that trans and non-trans men do, in fact, look, despite homophobic panic about gay male desire and effeminacy. There are ways of looking without being caught, and ways of seeing that fall under the radar. Haley, who is a non-trans queer femme, notices the paradox: 'We've established that men don't watch each other, but, um … they kind of do, right?' As one transmasculine interviewee who is active in campaigns for gender-neutral toilets hypothesizes, 'From what I understand … men do look [at each other at urinals] … you just don't be obvious about it … [this rule goes for] … straight men as well as gay men.' Nandita, who is bisexual and living in Toronto, says, 'My brother has sort of admitted to looking at every guy's penis.

Without their knowing ... and he would literally stand next to them and observe it ...' Gay, bisexual, and queer male interviewees also talk about how they violate prohibitions placed on queer optics. Tom, who is bisexual, admits to an unauthorized downward-pointing gaze at the urinal: 'I was peeking once ... leaned over [in a stadium urinal during a sporting event] ... and this straight guy [patrolling the bathroom yelled] "What the fuck you looking at faggot?" ' Ivan, who is queer, underscores how homosexual prohibitions engender fear:

> I think gay men are terrorized to not look at other men [in restrooms]. But ... I've been able to overcome a bit of that terror ... I feel vulnerable, but I'm forcing myself to look, and I'm developing a confidence in looking at men, even men I don't know are queer, or who are straight.

In restrooms in gay bars, mirrors are sometimes deliberately positioned so as to make the penis hyper-visible (thereby overriding homophobic prohibitions on looking at other men in the toilet). Zoe notes that there are gay bars

> where they have recently installed mirrors along the front of the urinals. And it's specifically for the purpose of checking one another's cocks out while using the urinal ... it's turning ... the mirror [around where it is no longer] ... being used for surveillance ... [and using it] in a really sort of playful, openly sexual way ... we [don't even have to] turn our heads, we can just check out everyone in line.

By sexualizing the floor space before the urinal, gay men who are not trans flout and undo the prohibitions placed upon homoerotic 'looks.' It must, however, be noted that looking at cocks at the urinal can be uncomfortable or dangerous for non- or pre-operative trans men. While not entirely trans-friendly, the eroticization and direct contravention of the prohibition placed on 'looking' at the urinal does promote homoerotism. What is normally a centre of heterosexist gender surveillance becomes a queer optic zone for non-trans gay men (as I discuss more fully in chapter 6).

The Stall and the Urinal

The restroom stall (like the urinal) is also a centre of panoptic surveillance. Foucault (1979) insists that if one is to understand disciplinary power it is crucial to take seriously its 'innumerable petty mechanisms ... [This will include] the subtle partitioning of individual behavior' (173) by design in panoptic space. Individuals are partitioned into enclosures (cubicles, primarily) that are individualizing and homogenizing in the present-day lavatory. There is what we might call the quarantine effect. As Foucault writes,

> Disciplinary space tends to be divided into as many sections as there are bodies or elements to be distributed. One must eliminate the effects of imprecise distributions, the uncontrolled disappearance of these individuals, their diffuse circulation, their unusable and dangerous coagulation. (Ibid., 143)

Without individualizing cells, desks, beds, and cubicles, the panoptic eye cannot function because it depends upon pedagogies of separation and demarcation. 'Visibility is a trap' (ibid., 200) set in space.

The partitioning of the body as it urinates and defecates is of crucial significance to the play of panoptic power. Cubicle walls and doors do not reach the floor. The gaps enable feet to be seen, while the person inside cannot know when or by whom or to what end a look may be given. Feet are subject to extensive visual scrutiny in the stall. Trans interviewees who use the cloistered closet in the 'women's' room lament the problem of surveillance below the knee. Patrons, who are usually non-trans (and not infrequently children), sometimes crouch down from adjoining stalls and from the common waiting area outside the cubicles to survey the gender of footwear, to monitor toe-direction, and to reflect critically upon what they perceive to be an unnecessary shuffling of the upper extremities. As JM, who is genderqueer recalls,

> A friend of mine [was] telling [me] how ... she had gone into the women's bathroom and some little kid had said 'Mommy she is facing the wrong way,' and then she went to the men's bathroom and somebody else had been like 'Daddy, he's facing the wrong way.'

Sugar, who is of Anglo-Scottish and Celtic heritage, living in Vancouver, and a non-trans queer femme, emphasizes that there can be 'confusion when a woman is standing the wrong way in front of a [toilet seat] because you can see someone's feet below the stall ...' Temperance, who is also a non-trans queer femme, confesses to partaking in the disciplinary apparatus by which gender is subject to surveillance:

> I look at feet underneath the stalls next to me, totally look at feet ... this is such a dirty confession, and I feel horrible ... But sometimes, I'll hear peeing next to me and I will look down to see if the toes are pointing outwards or inwards or I'll hear like a pee that sounds like it would only make that deep of a sound if they were standing from above and peeing down ... I think to myself ... what kind of shoes [are they wearing?] ... I totally police gender according to footwear from underneath the stall. Are those masculine shoes [or] are those feminine shoes?

The negative space beneath the stall is ideal for the promiscuous look. One cannot intercept the substratum gaze from outside the stall unless, of course, one crouches down below stall walls and cubicle doors – thereby risking allegations of sexual voyeurism and solicitation.

Some interviewees say they feel as though they are being watched, even when they cannot pinpoint the genesis of the look. As Isaac, who is a trans man, explains, 'I just – you just *know*. Like, you can feel people's eyes on you. And it's, it's something you get used to.' Jannie, who is a white, non-trans queer woman living in Toronto, illustrates the effect of the all-seeing eye in the stall:

> You can see people's feet, right, so ... when I go into the washroom ... the first thing that I do is put toilet paper on the seat, because I can't hover ... I have to actually sit on it [the toilet seat], so I cover it with toilet paper so my feet are facing the back [as I do this] ... I am thinking 'All right, everyone is looking at my feet. I look the way I look. They think that I'm a guy standing up.' So I ... [cover the seat with toilet paper] in [such] a way ... that my feet are pointing forward ... I turn my body [around as I] ... put the toilet paper on the seat.

The way the body is physically oriented and positioned in the stall ratifies, undoes, or incites questions about gender and genitals – so much so that Jannie undergoes rather extreme body contortions to control how her feet are seen by others, in adjoining stalls and in the common floor space outside.

The cubicle is, as suggested above, a disciplinary device authorizing a truth about the sex of the body.[3] The restroom gives people licence to survey and inspect gender and genitals in ways that are disallowed in other public venues. For many transgender, masculine, and butch interviewees using the 'women's' room, the act of leaving the stall and entering into the common floor space before the sink and mirror is angst-ridden and stress-inducing. Many say they are most likely to be subject to gender-based harassment when leaving the cubicle. As one trans interviewee notes, 'when I walk out of a stall it's worse than when I walk in.' Other people talk about waiting in the metallic compartment until it is safe to come out. KJ, drawing upon anecdotes from friends and lovers, says,

> I've heard of stories of people feeling like, they've gone into a stall and they've felt like a person outside of the stall was waiting around to see who's actually in [there]. Or they ... stay ... in the stall longer ... Waiting until other people came into the washroom so they felt like they could leave the stall [safely].

Callum says he often waits in the cubicle until his female partner comes to get him, so he doesn't have to stand on his own and 'brave the angry ladies.'[4]

Lana, who is a non-trans queer femme, discusses the problem of urinary positions and gender recognition in the 'men's' room:

> There's a lot of ... trans people [who] talk about going into the men's washroom and using the stalls instead of trying to use ... the urinals ... and then wondering if people will be able to tell if you're sitting down and peeing as opposed to standing up and peeing. Or, if they'll look underneath the thing and see which way your feet are turned ... there's all sorts of self-policing that comes into effect ... you're more aware [as a trans person] of what you're doing ... than other people may be.

Rico, who is a non-trans gay man, also says,

> I don't know if other guys do this, but if I go in the men's bathroom, and even if I go in the stall, I stand up. Because I'm cognizant that somebody might be looking at which way my feet are showing. And I don't like anybody to hear the toilet paper. And I always leave the seat up ... you don't want to leave any clues.

Jacq, who is genderqueer says, '[If] I'm going to use the male washroom, I'll go in and check to see if there's a stall because I don't use a urinal. If there is a stall, I'll check on the quality of the stall [whether or not the door locks and is secure], and then I'll [decide if I can] use it.' The concerns raised about stall doors indicate a larger worry about gender-based visual scrutiny – this despite prohibitions on 'looking' in the 'men's' room. One trans guy notes that

> In really small [men's] bathrooms ... where the stall door is either not there or it's really small ... People can see through, and I can't use those ones. Or I'd have to use them really fast and I'm totally terrified of my body getting noticed or identified ... It's just ... not wanting my body to get seen. Or like the consequences of what would happen, you know. Once your body is identified a certain way.

Callum elaborates on the problem of insecure cubicles: 'Often ... there aren't even locks on the doors ... I'm lucky if there's a door on a hinge ... it isn't an abnormal thing for there not to be a lock, or a door, or one that doesn't work ... It's very normal. In fact I'm more surprised if it locks than if it doesn't.' Even for transgender guys who use a funnel or stand-to-pee device (or stand to pee without them), stalls continue to be a necessity. As one interviewee who identifies as female-to-male transgender explains,

> I prefer a stall ... with the apparatus I have to use to pee standing up there is a bit of fishing around ... in crowd situations, it's impossible. You know during that [Toronto] Maple Leafs Game incident – I'm going to call it the Maple Leafs Game incident – I couldn't go, right. I'm standing at the urinal and my number came up and I got the urinal and I'm standing at the urinal and I couldn't go. I just put everything away and left.[5]

The Bodily Ego and the Bathroom Mirror

In her critique of the Foucauldian analytic of subjection in *Discipline and Punish*, Butler (1997b) says that there is a 'suppressed psychoanalysis in Foucault' (87). By this she means that he fails to account for how the disciplines generate subjects who do not always cohere or abide by its normalizing regimes. There is an under-theorized resistance to the disciplines in Foucault's work. She notes that the subject always exceeds the confines of his or her own disciplinary chains. Disciplines, like all prohibitions and taboos, are productive, and their effects cannot be purely repressive (as Foucault would agree) or ascertained in advance. There is no transparent or linear relationship between an idealizing norm, its deployment, and the way one assumes a relation to it. How one occupies the bathroom – abides by or refuses the body choreographies it authorizes – is less than straightforward. There must be a psychic, or libidinal, investment in a norm before it can be internalized or refused.

As Butler (ibid.) asks, 'To what extent is the body's site stabilized through a certain projective instability, one which Foucault cannot quite describe and which would perhaps engage him in the problematic of the ego as an imaginary function' (90)? If the disciplinary gaze plays a role in the functioning of the bodily ego, it is prudent to ask how glass- and societal-mirrors are central to gender identifications – to the way they are assembled and undone. The bodily ego is dependent upon a relationship (unstable and magnetic) between an internalized self-portrait (the genesis of which, Lacan says, originates in the 'mirror stage') and the non-visual sensational or proprioceptive ego (Silverman 1996) that takes note of one's location or position in space, feels its musculature shape and contours, its painful and pleasurable body sensations, and counts them all as his or her own. In her discussion of the bodily ego and materiality, Gayle Salamon (2004) suggests that the bodily ego is central to how the body manifests a sex. Mutually dependent upon visualization and corporeal sensations, the bodily ego (anchoring or providing an axis for gender) 'comes into being once the "literal body" assumes meaning through image, posture, and touch' (105–6).

To be out of spatial bounds in a gendered lavatory, for instance, is to be caught in a visual crossfire animating a lack of synergy, or synchronicity, between the body and its image. A normative cissexual boundary is one that overrides (negates, or refuses to avow) a gap between gender identification and the body corporeal as intercepted by others. A trans identification is often predicated upon an internal recognition of that gap and a desire to minimize or reduce it. In other words, the gap or fissure becomes integral to how one recognizes a discord between gender identity and the sex of the body.

Another way to conceive of the bodily ego is, in quintessentially Freudian vernacular, to imagine it as a psychic 'projection of a [bodily] surface.' This projection is, however, necessarily unstable and incomplete. Body-surface projections are sensorial and social. They are also psychically invested. In order to forge a relatively stable bodily ego, we use others as mirrors. We intercept the looks of others to consolidate our own internal self-portraits. Internalized projections of a bodily surface reverberate and unevenly touch upon (as they ignite and are shaped by) erotogenic zones. Genitals, orifices, and pain and pleasure receptors are all central to the formation of a bodily ego. As Silverman

(1996) tells us, the bodily ego is a 'stitching together' or composite of the visual imago (internalized self-portrait) and the sensational or perceptual body (18). The bodily ego (like gender) is the result of a psychically invested *process* (not an entity or possession of the self to be claimed as one's own). It can be undone and altered (as people change and grow), but most of us negotiate a relatively stable bodily ego in which the self takes shape.

The dynamic and unstable relation between self-image and sensation is what allows us to forge a self. The flexibility and movement between the two is crucial for ego functioning: coming 'undone' can be 'beneficial rather than tragic; it is the precondition for change, what must transpire if the ego is to form anew' (Silverman 1996, 21, quoting Schilder). The violent exposure and condemnation of the necessary gaps between images and sensations in the toilet is, however, painful and injurious, particularly for trans folk, who have less access to gender signifiers mirroring internal self-portraits in cissexist culture. Internal self-portraits do not fully match up with bodily contours, shapes, and morphologies 'as seen' by others; this is true for trans and non-trans people. Building upon the foundational work of Viennese psychoanalyst Paul Schilder, Kaja Silverman notes that the bodily ego lacks stability. 'Schilder emphasizes not only the tenuousness of the connection between the visual and the sensational egos but the looseness of the connection between bodily parts' (1996, 21). In essence, the internal self-portrait is in excess of the materiality of the body. The same point is also made by Didier Anzieu (1989), who writes that the 'Self does not necessarily coincide with the psychical apparatus; in many patients, parts of the body and/or the psyche are experienced as foreign' (19).

The visual imago may also disavow what is there to be seen. For example, a trans man may not 'feel,' avow, or invest in a 'breast' to be seen. The 'breast' as body part is not libidinalized. In this instance, the bodily ego is not shaped by the 'having' of a breast. It is shaped by the avowal of a masculine chest unseen by others unversed in trans-identifications. Schilder (1950) and Salamon contend that the '*body can and does exceed the confines of its own skin*. That is, the body is not an envelope for psyche, and the skin is not the envelope for the body: both body and psyche are characterized by their lability rather than their ability to contain' (Salamon 2004, 108). There is a 'lack' or 'excess' to the body that cannot be reduced to a corporeality (lacking libidinal investments and animation by the subject) or to that which is visually ratified by others. This 'mismatch' is particularly acute for transsexuals, who adopt internal body imagery that is often seen as incompatible with the morphologies of the body by those unschooled in trans embodiments and identifications.[6]

Trans people bear the burden of identifying with the distance between the visual imago and the sensational body – a distance that is normally disavowed by cissexuals. The mirror enables those who are able-bodied and cissexual to consolidate fantasies about what we might call an absolute juxtaposition of image and sensation. Those subject to heightened surveillance in bathroom mirrors are imposed upon to absorb a gap between an image and a libidinalized body for others who are not trans in ways that are inhumane. Such a projection 'encourages others [namely trans folk] to live the irreducibly disjunctive relation between the sensational ego and the specular imago in a pathological way, i.e., as personal insufficiency and failure' (Silverman 1996, 37). Gender normativity may be defined as a refusal to avow a gap in the self that is central to the bodily ego. This gap is often projected onto queer, physically disabled, and/or trans folk, who are, albeit differently, designated 'out of place' in the normative landscape or associated with volatile matter.

Distance from a degrading cultural screen is essential to a livable life for those whose identifications are inconsistent with normative imagery. 'The struggle here is not to *close* the distance between visual imago and the proprioceptive body, as in the classic account of identification, but to *maintain* it – to keep the screen ... at a safe remove from the sensational ego' (Silverman 1996, 28). Distance from the mirror – either reflective glass or a nullifying transphobic stare – can be essential to life. Insofar as transphobic looks circulate in the bathroom (and in other public institutions as well), it is necessary to refuse to carry the burden of identifying (negatively) with gaps that are foundational to the bodily ego. Distance taken from public mirrors is not pathological or predicated upon internalized transphobia. Nor is it something one needs to 'work on' or 'deal with' in a psychological sense (unless, of course, one wants to). Mirror avoidance (particularly in public bathrooms) is a viable and important strategy to refuse entry into transphobic 'mirror circuits' (Sedgwick 2003). By refusing the looking-glass, trans folk may also be demonstrating their entitlement to occupy gendered space without cissexual validation.

The mirrorical return can be injurious when it undermines the already fragile claim some trans folk have to their gender identities in cissexist culture. The visual echo conjures up what Silverman (1996) calls a lack of presence. 'The experience which each of us at times has of being "ourselves" – the triumph of what I have been designating the *moi* part of the bodily ego – depends on the smooth integration of the visual imago with the proprioceptive or sensational ego … When these two bodies come apart, that "presence" is lost' (17). 'Presence' (or a feeling of being 'at home' in one's body) is, for transsexuals, destabilized by the mirrorical returns in lavatories that amplify transphobic attentions to gaps, fissures, and disjunctures between identification and corporeality.[7]

Intersex folks and those identifying as trans often avoid the hanging glass because they dislike the images it reflects. Emily acknowledges a relationship between gender identity and mirror avoidance:

> I've avoided mirrors for many years. Because I dislike my appearance … I don't like going … anywhere where I have to look at myself … [But as] I'm more open with people [about my gender] … I've become more able to sort of look at myself in the mirror and be okay with what I'm seeing.

As Rohan, who is transmasculine, says,

> I don't feel entitled to stand in front of the mirror and check my look and fix my hair [as I do at home]. I want to get out of there [the public bathroom] as fast as I possibly fucking can … [This] … hyper … visible … space … in the bathroom, it's all about gender presentation.

Another interviewee explains:

> If you are already in a space … [where] you're otherized in, then the mirror, the existence of [the] mirror becomes even more of a way to police and make sure your body looks just the way it's supposed to look, right? … standing in front of a mirror … just [makes you] more vulnerable.

Because glass and societal mirrors project, they refuse to see. Probing eyes are transphobic and ableist, mirroring gaps, fissures, and disjunctions between the sexed body and gender identity, between the body as it is felt and the body as it is perceived by others. They also refuse to discern and to ratify modes of embodiment that are trans-generative, new, and based on what Susan Stryker (2006) calls a conscious 'disidentification' with 'compulsorily assigned subject positions' (253). As Jacq explains, 'Some people just pretend like [I am not here]… you're walking down the street or in a hallway, and they saw me, they would say hi ... I see them in the washroom and I say hi and they don't say anything.' Looking does not always enable one to see. Lana, who is from the American South, a white, working-class, non-trans queer woman, who uses a wheelchair, elaborates upon the ambivalence in the non-committal look in ableist and cissexist culture:

> I ... think [people] ... do it in that sort of I'm watching, but I'm not watching kind of way ... watching without looking at someone ... So, they might not be directly staring ... but they are listening, or like paying attention to where [a person seen is] ... going or what they're doing, and how they're doing it.

She also observes that 'People generally don't want to look at you, or ignore you ... you're not supposed to ever look at people with disabilities ... so ... people don't actually look at you directly, not very often ...' Phoebe, who is a trans woman (but not physically disabled), confirms that 'with a physical disability, there's visible, and then there's invisible.' Madison, who has worked with people with disabilities, believes that there are social codes discouraging direct looks, but that bathroom designs are set up for visual scrutiny:

> When somebody's ... in a wheelchair or has a mobility issue ... people don't want to stare, because they don't want to be rude, but the architecture of bathrooms lends itself to [the] look when that person doesn't know that you're looking. For example, the stalls don't go all the way to the floor, the mirrors, or even ... [when] ... the stall isn't big enough for ... [someone with a mobility issue and an] ... attendant, so often the door will ... be open ... the architecture lends itself to sneaking a peek. But [in] keeping with these ... rules or ideas about ... not looking or making eye contact ... people can look away, but [they do] check out how people in wheelchairs use bathrooms.

Madison also notices how people with visible physical disabilities are subject to public voyeurism. She suggests that able-bodied cissexuals are often intrigued by the gender, genitals, and urinary capacities of the disabled:

> [If a person with disabilities] ... comes into the bathroom ... people certainly watch them to see what they do ... if somebody comes in with a

> cane, or a wheelchair, and they're, like, obviously coming to use the bathroom ... people watch them go into their stall, and they watch them to see how easy it is for them to open or close the stall door. Or people may watch ... their feet to see how long it takes them to get on the toilet, or if they can reach the sink ... I worked with a woman who had one arm and ... when we were in the bathroom everyone was, like, so how are you going to wash your hand?

Personal-care assistants accompanying people with physical disabilities often attract voyeuristic attention when both parties occupy the same cubicle. Onlookers wonder what might be happening in the stall. How genitals are touched and wiped, how bodies are held or lifted up and over the porcelain toilet seat by assistants are all points of intrigue. Technology – wheelchairs, crutches, and other facilitating devices – heighten the interest of onlookers, who may also be troubled and undone by what they see (of themselves) in the faces of disabled people.

As Michael Davidson (2003) notes in his discussion of how people with physical disabilities are represented in film noir to enable a heteronormative plot line, the disabled are seen to be volatile; they are 'volatile because they make visible the field of sexuality itself, not as a set of drives toward an object but as an array of positions, desires, acts, and practices' (71). Noting a common set of cinematic techniques and devices through which both queer and disabled characters are deployed to signify abnormality and perversity, Davidson further suggests that the presence of a disabled character produces a 'triangular gaze among viewer, noir hero and femme fatale' (60) that is seen to undermine and to interrupt normative heterosexuality dependent upon the filming of able-bodied and conventionally gendered protagonists – leading 'ladies' and 'men.' But the presence of the disabled also enables a heteronormative resolution by absorbing or standing in for a queer or perverse remainder (or excess). The disabled are deployed as a narrative crutch or as a 'narrative prosthetic device' (Mitchell and Snyder 2000, 49) to quell anxiety about gender disorder.

Lana explains how her personal-care assistant, who is a transgender man, is typically not harassed when he is with her in the bathroom:

> I see how my [cissexual] presence [makes my personal-care assistant's] gender ... [eligible] to be in that bathroom ... a friend of mine who's trans and who ... generally ... has difficulty negotiating bathrooms because they're not seen as really fitting in either space ... if he's with me, then it's

okay for him to be in whatever bathroom, and he never gets any harassment ... So ... even if I don't have to go to the bathroom, I'll go with him, so that he then would have, like, a legitimate reason for then being in the bathroom, and nobody would question him ... Plus, he's seen as a good person because he's helping me out, right?

Not only does Lana 'enable' entry into the toilet, but she offsets cissexist worries about gender-based encroachments in public space.

Multiple sets of eyes become an issue, however, when crowds of people who do not know, and may not confer humane gender recognitions, wait outside the door. Many interviewees agonize about standing alone in bathroom line-ups. Temperance, for example, refers to bathroom lines as 'visual parades.' A few trans interviewees do, however, say that the concession can neutralize what would otherwise be an antagonistic bathroom encounter because it gives people in line time to register their presence. Sarah says, 'I like it better if I'm waiting in line. It gives people ... an opportunity to, like, size me up and figure out what's going on.' Callum elaborates:

Everything that happens in a ['women's'] washroom ... happens to you when you don't look a certain way, you look male, so it's all that badness, but prolonged and concentrated [in a line-up] ... when you're standing in line ... there's a group sort of standing there looking at you ... as soon as I enter *into* the bathroom, suddenly it becomes ... a problem [in a way that it isn't when I am standing in the back of the line-up in the hallway] ... [the] freak-out that's happening ... [doesn't take place in the] ... hallway ... but once you get *into* the washroom it [my presence] ... becomes ... [subject to] ... overwhelming panic ... there have [also] been times when ... I can stand in the line ... [and] get to know the women that you're going to be in the washroom with ... [they] ... get ... used to the idea that ... [I am] going into the washroom ... [and] ... I can do my 'tits out, voice high' thing [to accentuate a female form to] prepare them.

Trans folk, as indicated by Callum, use multiple strategies in line-ups and in the 'women's' room to navigate cissexist mirror circuits. One interviewee talks about her friend who uses 'jewellery ... that she keeps ... under her shirt ... and ... whips out [when she enters the bathroom].' As Callum notes, 'You just change your posture so that you [adopt] ... a gendered, stereotyped way that women stand or carry themselves.' Voices are also used to accentuate a feminine tone (usually at odds with one's usual or preferred intonation). Isaac says, 'I put my voice ... up like this [his pitch rises] and talk to her [his girlfriend] like this [voice getting higher in pitch], and stick my, stick my boobs out, and I'll talk to her the whole time.' Rocky explains how he 'will somehow clear my throat and walk in a certain posture and make sure I feminize some aspect that's notable.'

Trans folk sometimes parody normative gender scripts to gain access to gender exclusionary space. Linda Hutcheon (2000) defines parody as a 'form of repetition with ironic critical distance, marking difference rather than similarity ... [There is a] tension between the potentially conservative effect of repetition and the potentially revolutionary impact of difference ...' (xii). The effects of parodic performances are dialectical and uneven. In other words, there is a 'double-voiced discourse' (ibid., xiv), one with an ironic and critical distance from the regulatory norm or idea, but also a conciliatory or partial recognition of the norm. While a gender parody is, for some, an ineffectual or limiting strategy, it is for others a necessary strategy. Gender parodies are seen by cissexuals to animate (libidinize) the flesh. By staging, or rather parodying, a synchronization (between gender identity and sex) that the normative cissexual audience wants to see, the trans subject returns (through ironic and critical distancing) normative investments in symmetry. But there is something changed (asymmetrical or altered) in the mirrorical return. The normative accord between gender identity and sexed embodiment is uprooted because it is seen by non-trans people to be less than fitting, or perhaps at odds with the performer's internal mirrors (self-imagery) and related gender identifications. In the following chapter, I illustrate how the lavatory disciplines gender by deploying what Kaja Silverman (1998) calls the acoustic mirror.

4 Hearing Gender: Acoustic Mirrors[1] – Vocal and Urinary Dis/Symmetries

> [S]ometimes the disjunction between the visible and the verbal may point to meanings that lie between them.
>
> (Marks 2000, 129)

> There is here a dual dialectic of the senses, of seeing and hearing: no one is to be allowed to see me ... sound also violates, and submission to other people's sounds is a symbolic index to powerlessness and vulnerability.
>
> (Leach 1997, 264)

> Freud's work on paranoia ... suggests that vision and hearing play a key role in the relocation of an unwanted quality from the inside [of the body] to the outside ... Visual and auditory hallucinations have a critically important projective function.
>
> (Silverman 1988, 16)

Disciplinary power depends upon visual surveillance, but the surveying of bodies in modern institutions has an auditory component as well. It is not a well-known fact that Jeremy Bentham entertained the idea of developing an acoustic surveillance system in his modern prisons.

> In his first version of the Panopticon, Bentham had also imagined an acoustic surveillance, operated by means of pipes leading from the cells to the central tower ... [But the idea was abandoned] because he could not introduce into it the principle of dissymmetry and prevent the prisoners from hearing the inspector as well as the inspector hearing them. (Leach 1997, note 4)

The sounds of prisoners and prison guards echoing down a pipeline would not be one way but two way. Both the inmate and the guard would be intercepted by the omnidirectional sound waves. Neil Leach (1997) hypothesizes, in a footnote, that the symmetry of the acoustic vibration (I hear you, as you hear me) could not be overcome by the introduction of a dissymetrical pipeline, and so an acoustic surveillance system did not become part of the panopticon proper. The flow and traffic of sound couldn't be routed to obscure the location of the one who hears. Acoustic mirrors do not operate like one-way glass mirrors, cameras, observation towers, and other visual technologies that obscure the seeing/being seen dyad. Sounds vibrate, echo, carry, and reverberate in ways that are less amenable to unidirectional silencing. Sounds are more diffuse and unruly than lines of vision.

In modern Western architectural design there is, as critical geographers note, an unprecedented focus on optics and acoustics.[2] Public bathrooms, like other modern institutions – prisons, hospitals, scientific labs, psychiatric asylums, and so on – amplify sound. Silence is to order or acoustic purity as noise is to disorder and acoustic impurity in the disciplinary institution. Peter Bailey defines noise as 'sound out of place' (quoted in Cockayne 2007, 113). 'Sounds are out of place when issued in an inappropriate place, or at inappropriate times' (ibid.), thereby upsetting the predictability of the acoustic tempo – its genesis, radius, or point of origin. Senses are most often offended when they are in contradistinction to expectations. Emily Cockayne argues that in premodern England the 'threshold of decency changed' (ibid., 231), and this shift in auditory sensibilities leading up to the present day is especially evident in the gendering of elimination. If there is anything universal or unisex about the body it is in its sounding of elimination. The sounds are offensive because they confuse gender. The gendering of the lavatory by design ensures order or acoustic purity where there would otherwise be none.

Acoustic mirrors are forms of entrapment. Our auditory capacity enables us not only to apprehend shapes and surfaces but also to hear others. 'With the echo the sense of distance as well as surface is present. And again surface significations anticipate the hearing of interiors' (Ihde 2007, 68). There are auditory spatial significations to the way we use and inhabit the lavatory. Through the ear, we apprehend human interiors that are normally invisible to the naked eye. 'Hearing interiors is part of the ordinary signification of sound presence and is ordinarily employed when one wishes to penetrate the invisible' (ibid., 71). If a picture is worth a thousand words, voice must also be said to generate, and to work in harmony with, the pictorial. The human ear is 'given the imaginary power to place not only sounds but meaning in the here and now ... it is understood as closing the gap between signifier and signified ... Western metaphysics has fostered the illusion that speech is able to express the speaker's inner essence' (Silverman 1988, 43). Sounds can also highlight what is taken to be a lack of presence, a gap or fissure between what is felt on the inside and what is seen or heard on the outside. An acoustic can signify what is present *as well as what is missing*.

Acoustic architectures in public lavatories are cold, clinical, and sterile; not warm, inviting, and choric. White tiles, metallic walls and taps, porcelain toilets and urinals, glass or metal mirrors, right angles, smooth and flat surfaces all echo gender. Genitals normally concealed from sight are sounded in the lavatory. Interiors are made present by acoustic designs. Acoustic mirrors in the modern lavatory amplify interiority, and no one is exempt from the echoic exchange. Because the 'voice is capable of being internalized at the same time as it is externalized, it can spill over from subject to object and object to subject, violating the bodily limits upon which classic subjectivity depends, and so smoothing the way for projection and introjection' (Silverman 1988, 80). Acoustic mirrors, be they human ears or more passive acoustic objects like tiles and metallic surfaces enabling the rebound and play of sound in the lavatory, are used to apprehend gender.

Acoustic mirrors work in harmony with the visual to discipline gender. While Kaja Silverman's Lacanian analysis of acoustic mirrors is focused upon cinema, there are important parallels between the film as a visual production enabling the audience to look while not being seen and the toilet as a modern room where one looks and hopes not to be seen. The cinematic voyeur and the bathroom patron occupy a liminal space obscuring the centre of the mirrorical gaze and probing ear. Silverman (1988) argues that classic cinema orchestrates a division between 'the site of enunciation and that of spectatorship' (180), between interiority and exteriority, and obliges the feminine subject to carry the burden of lack disavowed by the masculine subject. The burden is transferred onto the feminine subject by visual and acoustic mirrors. The modern lavatory sets visual and acoustic mirrors into play. It is designed to align feminine people with castration, (loss, subjection, and interiority),[3] and to align the masculine subject with phallic potency (exteriority, objectivity, and autonomy). It should also be noted that Silverman's most vivid illustrations of acoustic mirrors involve bathroom scenes in films like Frances Ford Coppola's *The Conversation* (1974), Alfred Hitchcock's *Psycho* (1960), and Michael Powell's *Peeping Tom* (1960). For example, in her discussion of how the female voice functions as an acoustic mirror for the male protagonist in *The Conversation*, Silverman analyses how the hero (Gene Hackman, who plays Harry Caul) repeatedly returns to the toilet, which is symbolically coded as sonorous envelope and place of subject non-differentiation in utero, to wrestle with his dual need for autonomy (discursive control) and pre-symbolic wholeness to be found in the echo of the mother's voice in the sonorous envelope.[4]

The toilet is designed to absorb and to contain the spectre of loss and impotence for the masculine subject. The feminine subject – whether she be trans or cissexual – is coded as carrier of loss and impotence: her genitals are invisible (cloistered in stalls) as if there is 'nothing to see' (or, conversely, something to hide) while masculine genitals are hypervisible (at the urinal). 'The mirror also functions to disavow male lack even further by suggesting that the female subject reflects only herself, and so denying the place of the male subject within the mimetic circuit' (Silverman 1988, 34). Mirrors are like envelopes (Rose 1996, 70); they highlight contours and edges. The mirror intercepts and contains the subject. But our relation to form and matter, to solids and liquids, is gendered and gendering. The feminine is characterized by formlessness and a radical indeterminacy. 'The feminine other remains the inexpressible embodiment of alterity … she is a negativity which circulates towards but never returns to herself as a locus of development of a positive form' (Vasseleu 1998, 124–5). The masculine, by contrast, is fully severed from the mother, or so he appears and pretends to be. The feminine reveals the residual trace of the mother. She is porous, leaky, and ill-defined. This archetypal construction of the feminine in the Western cultural imagination impacts upon the way women (trans and cissexual alike) are viewed as having a more fragile claim to an independent and autonomous subject position.

While trans men lay claims to masculinity, their identifications with it may be less often predicated upon an absolute severance from the mother (or the archetype of the maternal-feminine). Of course, some trans guys may, like some cissexual men, equate feminine people with the spectres of loss, interiority, and castration disavowed in themselves. Some trans men may transfer the burden of identifying with difference onto feminine people as compulsively as do some heteronormative and cissexual men. It may be true that some trans men wish to consolidate the master signifiers of sexual difference (male and female, masculine and feminine) because they are the markers used to navigate and authenticate what Jay Prosser (1998b) calls the transsexual trajectory. It should also be emphasized that transsexuals are often compelled to foster a belief in binary gender codes to access medical intervention and to get hormone prescriptions. Of course, trans men (like trans women) may also be more willing to avow a space in between the two master signifiers of sexual difference because they are acutely aware of how the medical and psychiatric establishment uses cissexist gender codes to limit access to sex-reassignment surgeries.

It is, however, clear that trans men interviewed in the *Queering Bathrooms* study are more likely to avow their status as castrated subjects, their relation to loss, and to be critical of how feminine and feminized people are aligned with abjection. While the feminine is often a symbol or embodiment of castration (lack, subjection, and loss) for the non-trans heteromasculine subject, trans folk lacking access to cissexual privilege are also – albeit in different ways – culturally aligned with a material abyss or a lack of concrete form, interiority, and subject instability. This alignment produces what trans theorists refer to as gender incoherence or, rather, as 'illegible' bodies. I suggest that transphobic refusals to see the body cohere outside of normative cissexual bounds parallel (and work alongside) a related process of feminization (theorized by Silverman). Those who are recognizably trans are imposed upon to embody a gap or disjuncture between image and sound (that is feminizing because it undermines subject autonomy and cohesion) by non-trans folk, who are more likely to demand an absolute accord between what can be seen and heard in heteronormative and cissexist mirror circuits. What Silverman (1988) calls 'verbal and auditory defences' (41), along with visual defences, work in harmony to fortify subjectivity. Gender stasis is consolidated through an illusory unison, and by objectification. 'Graphically, the "individual" might be pictured as a closed circle: its smooth contours ensure its clear division from its location, as well as assuring its internal coherence and consistency' (Kirby 1996, 45).

Gender stasis, its reversibility, and its indeterminate points of origin are at issue when sounds and images are out of sync. Worries about acoustic purity (silence), and related worries about the genesis and direction of sound energies, are manifestations of deep-seated concerns about how the subject will be intercepted by acoustic mirrors. As Steven Connor (1997a) notes in his discussion of noise and affect,

'Everywhere the question seems to be how to measure, predict, regulate and limit noise, rather than to understand the complex interchanges between desire and endangerment involved in noise' (154). David Trotter, building upon Connor's seminal work (1997b), notes that 'sound, unlike sight, has often been understood as a disintegrative principle. We can only see one thing at a time, but it is possible to hear several things simultaneously. Sound often carries menace unless and until we trace it back to and locate it in a specific source, or visualize its origin' (Trotter 2005, 37). This is why the acoustic design of public lavatories enables questioning about genitals and interiority, the presumed 'origins' or determining features of gender. Bathrooms closely resemble intrauterine maternal spaces; receptacles for the subject as yet unsevered and not discursively defined. The lavatory is felt to be strangely anachronistic for the adult occupant, out of place and time: 'strange and antiquated,' as one interviewee notes. An echo in the toilet is temporal and uncanny, familiar and strange. Because the echo defies capture, it incites questions about its origins, the object it signifies and circumscribes by vibration, along with the object's temporality and positioning in space. Despite the best efforts of acoustic architects to offset the intimacy (or homey feel) of the public restroom, it continues to remind us of interiority and of what Lacan calls the imaginary (a pre-linguistic image-based phase in human development). The interiority of the room upsets our certainties about the unitary 'I' – or, rather, the plumbing and metaphysics of gender deployed to secure our status as subjects in the social field. While sound can work in harmony with 'sight' to apprehend gender in the toilet, it can also undermine our certainties about gender. The cissexist designs of the toilet are built upon a wish for normative gender synchronicities measured and authorized by visual and acoustic sensory registers – externalizing senses, not tactile or olfactory registers that are more intimate and less objectifying.

Gender and the Acoustic Mirror

Acoustic designs are not neutral or passive. 'An aural architect can create a space that encourages or discourages social cohesion among its inhabitants' (Blesser and Salter 2007, 5). Aural architectures in the toilet are individualizing and rigidly gendered. Sight and sound work in concord. We learn to 'see' gender with our ears. 'Knowing by seeing and knowing by speaking/hearing are gendered, respectively, as masculine and feminine ...' (Stryker 2006, 247).[5] The ear ascertains what is often hidden from sight. We visualize objects by acoustic cues. Public facilities amplify, as they also conceal, a lack of concordance between the body seen and the body heard. Lack of synchronization between body sounds and body images puts gender at risk. Interviewees agonize about voice and urinary acoustics authenticating or invalidating gender. Genitals are inferred by sound when they cannot be seen. The pitch and range of vocal chords, the trickle of urine, the shuffle of feet in a neighbouring stall, all spark curiosities about genital composition. Urinary positions are contrived so as to accentuate and to din tintinnabulation.

Those who are trans cannot always abide by the auditory and visual synchronizations used to consolidate normative male and female gender identities. Interviewees strive to emulate, or to mimic, the co-ordinated patterning upon which cissexist bifurcations of gender are based. But they also undo and trouble the equilibrium. By exposing the artifice, the precarious and constructed nature of the concordance between sound and image, gender non-conforming and trans folk defy (as they also mimic and parody) heteronormative and cissexist rules of symmetry, by choice or by default. Disunities between sight and sound signify something in excess of the visual. This excess denotes the double-sidedness of the human subject intercepted at a limit or visual horizon. 'The horizon is that most extreme and implicit fringe of experience that stands in constant ratio to the "easy presence" of central focusing' (Ihde 2007, 108). The vanishing point (or plumb line) defies representation in visual and acoustic fields and signifies the indeterminacy, or perhaps the liquidity, of the gendered subject.

Heteronormative and cissexist synchronicities demand a concordance between body sounds and images in the case of the feminine, whereas a greater tolerance for discord is evident in the case of the masculine. Kaja Silverman (1988) notices how in Hollywood film there is a 'close identification of the female voice with spectacle and the body, and a certain aspiration of the male voice to invisibility and anonymity ... Hollywood pits the disembodied male voice against the synchronized female voice' (39). The male voice is omnipotent and disconnected from its corporeal coordinates where flesh can be severed or hurt. The female voice signifies vulnerability (symbolic castration) and makes the body visually apprehensible. It is situated at the point of textual origins. The masculine voice is authorial, while the feminine voice is embodied and divested of agency. The female voice does not easily (or often) escape its bodily encasing or cultural alignment with castration. The male voice, by contrast, is under no obligation to synchronize itself with a visible corporeal encasing.

Sounds are associated with feelings of loss, vulnerability, and a lack of discernment between self and other, subject and object. 'Because we hear before we see, the voice is also closely identified with the infantile scene … it is through the voice that the subject normally accedes to language, and thereby sacrifices its life; it is associated as well with phenomenal loss, the birth of desire, and the aspiration toward discursive mastery' (Silverman 1988, 44). The voice, like the eye, is a means to project and to disavow castration. There is a projective defence employed in Hollywood film, much as there is in bathroom acoustic designs. Women are aligned with interiority in film: 'Interiority in Hollywood films implies linguistic constraint and physical confinement – confinement to the body, to claustral spaces, and to inner narratives' (ibid., 45). In the toilet, the feminine is similarly confined to an architectural interiority (the stall), while the masculine is displayed before an open-concept urinal. Female genitalia are rendered invisible (nothing to see), as they are also visual emblems of castration. Male genitalia are hyper-visible, but not to be subjected to visual scrutiny. Feminine subjects are, by contrast, repositories of lack. Paradoxically, the lack must be made evident or signified by an acoustic apprehending the body in space.

Phallic economies demand that the 'woman' be a 'little man.' The female clitoris is either seen to be an inferior version of the penis, or female genitals are seen to be a complement to the 'penis, in which case, it is the vagina which becomes the focus of attention. When woman is not held to one of these two representations, she is obliged to function as a stand-in for the male subject's mother, who is the only (heterosexual) object he is ever capable of loving' (Silverman 1988, 143). In other words, heteronormative and cissexist (or perhaps phallocentric) laws of symmetry negate sexual difference as they also insist upon female genital inferiority (clitoris as 'little penis'). This is related to what Luce Irigaray observes in her critical analysis of the unsexed and universalist laws of objective reasoning, homogeneity, and the associated masculinist fantasy of sexual sameness (negation of the feminine) in modern architectural design. The sexed subject (who is usually a cissexual woman in Irigaray's account) is not only negated by modern design but actively incorporated into a unisex or allegedly gender-neutral space of non-difference. For Irigaray, gender or sexed neutrality (spatial and otherwise) signals a masculine negation of feminine difference. In other words, feminine and, I would add, trans difference does and does not exist.

The enigma is built into bathroom engineering and plumbing designs. The urinal, for example, is a larger or more publicly visible receptacle amplifying the presence (or absence) of masculine genital organs (requiring people to stand upright and forward facing), while the oval toilet bowl (meant for sitting and backing into) is cloistered by enclosed cubicles. Furthermore, the mirror – more prominently displayed and less likely to be broken in the 'women's' room – aligns users with an objectifying gaze; patrons are forever reminded that they are spectacles or objects to be seen. Public toilets are at once segregated (insisting upon an absolute difference between the sexes) and strangely the same (urinals and amplified glass mirrors notwithstanding). There is an uncanny likeness between the two rooms. Most people have, at one time or another, walked into the 'wrong' bathroom and not known it right away. There is a pregnant moment in which we are uncertain about sexual difference and where we fit in relation to it.

Heteronormative (or phallic) symmetries insist upon seeing 'woman' as an incomplete or inferior copy of 'man,' upon a disavowed homosexuality, and also upon a splitting of female bodies into Virgin – (*object (a)*)[6] – or Whore – (abject) – dichotomies. So how do trans folk register in such an economy? I suggest that, not unlike the female genitals, which are and are not like the penis, hidden and ultra-visible (if only by acoustic projections in space), trans folk are invisible and hyper-visible at the same time. While the feminine subject is employed to contain loss for the non-trans masculine subject in Hollywood film, and in public facilities more generally, recognizably trans folk – transmasculine and transfeminine alike, along with those who are gender non-conforming – are imposed upon to absorb the asymmetries of the bodily ego (the discord between the visual imago and the sensational ego) that are often disavowed by conventionally gendered cissexuals.

Those who are recognizably trans and/or gender non-conforming upset the master signifiers of sexual difference that heteronormative and non-trans people often depend upon. Even when transsexuals invest in their identities *as women* or *as men* unambiguously, and strive to secure those same master signifiers of sexual difference (in themselves, in others, and in society at large), their status *as trans* (when apprehended by sight and sound) upsets the essential fixity of the sexed body, revealing it to be something that can be altered and brought into alignment with a psychically invested gender identity. Those invested in transphobic mirror circuits are troubled by the extent to which transsexuals lay claim to what is felt on the inside, as opposed to what is to be seen or heard on the outside. The privileging of one's inner psychic life and identificatory investments over the body's corporeal shape and genital form as intercepted by others, by sight and sound, violates an essentialist and transphobic law of sexual difference that depends upon a disavowal of the former and an avowal of the latter.

The spatial signifiers of sexual difference are, as I argue in what follows, troubled by sound. Vocal chords transcend the body and upset fantasies about a transcendental metaphysics of substance, an inner bodily core upon which sexual difference, its purity and legibility, is concretized. The rules and regulations governing voice in the lavatory are a case in point. Gender purity is established by voice. The feminine sounds the body, whereas the masculine silences the body. The moratorium on speech in the 'men's' room is often emphasized by interviewees. According to Gypsey, who is transgenderist (and living full-time as female), 'I know from my guy days that I never talked to anybody in the bathroom. I went in, I did my thing and I left.' An interviewee who is bisexual observes how speech bans are most likely to be in effect at the urinal:

> I have seen two men who are clearly friends come in [to the bathroom], they are having a conversation as they walk through the door and then they open up their pants and start peeing and the conversation stops. And then as soon as the pants are zipped back up again the conversation starts again in the exact same place as if it just hit a pause button.

Emily, who is intersex, notices how prohibitions on talk in the 'men's' room produce a dismal and isolating environment: 'Men won't even speak to each other to ... [exchange basic] information [about a broken hand-dryer, for example] ... It's ... very individualistic, very silent. Very sepulchral, almost.' David, who is a non-trans gay man, confirms that the 'men's' room is 'not the place to start conversation.'

By chatting in the 'men's' toilet one will likely be perceived as gay or as effeminate. David offers a detailed synopsis:

> If you are chatting with each other ... [it] could ... be constructed as picking up or expressing homosexual interest ... I am ... more identifiable as gay when I open my mouth than when I am just using the washroom. So perhaps because it tends to be a very heterocentric ... space, it's good for me not to talk to anybody and by talking I would often end up implicitly disclosing my sexual orientation.

Neil, who is a trans man and uses the 'men's' toilet, says that this room is an undeclared homo-social space. Chatter is forbidden because it accentuates the otherwise disavowed 'gayness' of the place: '[The] guy's bathroom, it's got too much of that latent homosexual appeal, it's too queer-as-folk, I guess, in the hetero mind … It's just a big no-no [to chat].' Not coincidentally, the prohibitions on speech are not often enforced in gay bars. Emily notes the double code: 'Men, as a rule … are not likely to speak to one another [in bathrooms but] … in gay bars, I find that gay men chatter endlessly in the bathrooms with one another.' Rohan believes that refusals to adhere to speech bans are very much about the queering of masculine toilet space: 'Queer men in a queer event with queer men, they look at each other, they talk to each other they cruise each other, they fuck each other.'

The government of speech is gender specific. In the 'women's' room, chatter is almost an imperative. While the 'men's' room is described as 'individualistic,' 'sepulchral,' and 'anti-social,' the 'women's' room is said to be 'communal,' 'lively,' and 'social.' According to Neil,

> You walk in [to the 'women's' room] and for some reason, it's like walking into a club where you are automatically assumed to be one of them and then someone … randomly strike[s] up a conversation about how they like the marble that the sink is made of or whatever.

Rohan is similarly amused by the sociality of the 'women's' room:

> Many women … speak to strangers … they are like 'Oh, how are you doing?' and … 'Oh, I like that lipstick,' and 'Oh, I got it here' [while another is saying] 'God, I am so pissed off.' A woman will just strike up a conversation with an … utter stranger in a washroom … there is … this sort of community thing that happens.

The feminine toilet is a social space, but this sociability is exclusive. Those who depart from feminine gender norms are often excluded. For example, as Rohan explains, 'I have never felt a part of that [bathroom culture] … I am not included as part of that community.' The exclusivity is a problem because talk is often a prerequisite for entry. If one does not talk or engage in social 'chatter' there is something thought to be 'up with your gender,' as one interviewee says.

There is, however, no incitement to talk between cubicles. As Sugar, who is a non-trans queer femme, notes, 'You have to know someone pretty well before you decide that you're going to talk between stalls. Versus in the common mirror of the washroom.' The space before the sink and mirror is where talk between patrons most often occurs. Talk between partitions is far more likely to be guarded or circumspect. The relative embargo on speech between cubicles reveals a wish to avoid a lack of synchronicity where one can be heard but not seen. There is an ever-present worry about a lack of symmetry between the feminine voice and its point of genesis. Vocal acoustics must be heard and seen in the place of the feminine body (its point of origin), and in the time of enunciation. Acoustic vibrations obscuring points of origin upset femininity because there is a demand placed upon female bodies to synchronize voice and image. While it is not uncommon for a male voice to radiate in ways that obscure its point of enunciation, female voices are imposed upon to speak in the presence of a body that can be visually apprehended. Feminine voices must be anchored in visible bodies. An embodied voice is feminine, whereas a disembodied voice, authoritative and territorial, is masculine. Sounds heard out of sync and out of alignment with the body trouble feminine gender identifications.

Masculinity, by contrast, has a right to emit and to spill and to take up acoustic space; femininity is to be fluid, bound, and subject to constraint. In Hollywood film it is not uncommon to confine female voices to a 'recessed area of diegesis' (Silverman 1988, 61), an interiority or space of confinement through which loss, shame, or castration can be seen to reside in the feminine subject, as opposed to the masculine subject. 'The identification of the female voice with the female body thus returns us definitively to the scene of castration' (ibid., 62). Loss is signified by one's proximity to the body and to its interior space.

In his discussion of the phenomenology of sound, Don Ihde (2007) suggests that sound differs from sight because it penetrates the body. 'In terms of sound penetrability, however, the escape from or control of sound is essentially a matter of psychic control' (82). In other words, sounds can be both heard and 'unheard.' User's of the 'women's' room tend to be troubled by sound vibrations between stalls. The trouble prompts one not to hear what is there to be heard. Excremental sounds (and flatulence) are often 'unheard.' Voices that obscure the visual points of enunciation are also wilfully ignored. Madison, who is genderqueer comments on the performative unhearing:

> You're supposed to pretend to not hear anything [in the 'women's' room] ... you're not supposed to hear ... [people] ... in their stall doing anything, whether it's changing their pad, or ... whatever ... you're just not supposed to hear.

As Rohan speculates, 'I think women are socialized not to talk about toilets or excrement or urine or menstruation or sex, sex in bathrooms, anything to do with genitals, body waste, women are socialized [not] to talk about that.' Another interviewee, who is queer, notes that those who use the 'women's' room are 'praying that nobody else is here [in the bathroom] or … really embarrassed about the [sounds of urine and faeces] … It just sounds like the person in the stall is embarrassed to be making noises.' The same interviewee notices that '[bathrooms] … are designed to be easy to clean, which … makes them loud and echoing and … there is almost a furtive quality to it.' A feminine echo must be apprehended in visual range of the body. Feminine sounds must be heard in visual range of the body so as to reveal (or 'out') the insides, one's spatial and temporal coordinates.[7]

As disciplinary institutions, toilets are designed to make gender anomalies visible and noisy. While feminine patrons often endeavour to synchronize the visual and the acoustic, the bathroom disorders images and sounds. Sounds cannot be easily apprehended at their points of origin. One never knows for sure from which stall a voice (or excremental echo) radiates, or from which urinal a sound vibrates (at least not without a sideways glance). Nor do we know who hears our own emissions, from where, or to what end. Images are not always in keeping with acoustic mirrors. Speaking about the maternal voice as sonorous envelope (and how it is used in Hollywood film), Silverman writes of 'the double organization of the vocal/auditory system, which permits a speaker to function at the same time as listener, his or her voice returning as sound in the process of utterance. The simultaneity of these two actions makes it difficult to situate the voice, to know whether it is "outside" or "inside." The boundary separating exteriority from interiority is blurred by this aural undecidability – by the replication within the former arena of something which seems to have its inception in the latter' (1988, 79).

The acoustic mirror is used to discern self and other, but it is also a means through which those same delineations are confused and undone. Silverman argues that 'whereas the mother's voice initially functions as the acoustic mirror in which the child discovers its identity and voice, it later functions as the acoustic mirror in which the male subject hears all the repudiated elements of his infantile babble' (1988, 81). The hearing is dangerous for the adult heteromasculine and non-trans subject because feminine speech (coded as babble) reminds him of his own relation to loss, castration, and interiority, qualities that are typically projected onto feminine people. Silverman is here presuming a non-trans male gender identity, and her reference to the feminine is quintessentially cissexual. It remains to be seen if trans men are upset by feminine (maternal) speech in ways that are comparable to those normally attributed to cissexual men. If white, heteronormative masculinities (trans and non-trans) are dependent upon the repudiation of 'female babble,' and of femininity more generally, it may be that those who are transmasculine, butches, and trans men do have a wish to locate and to quarantine the female voice. However, others may not need to transfer their own fear of engulfment or castration anxiety onto feminine people.

Silverman suggests that the 'very reversibility which facilitates these introjections and projections also threatens to undermine them – to reappropriate from the male subject what he has incorporated, or to return to him what he has thrown away' (1988, 81). The doubleness of voice gives gender away. 'The very possibility of an essentially "doubled" voice is a possibility that holds that every "expression" also hides something that remains hidden and thus cannot be made "pure" ' (Ihde 2007, 179). The acoustic politics of the toilet endeavour to separate masculine and feminine voices, interiority and exteriority. By sounding confusion (or penetrability) between the two, acoustic mirrors reveal less than coherent subjects. Paraphrasing the formative work of Guy Rosolato (1974), Silverman further suggests that, 'Since the voice is capable of being internalized at the same time as it is externalized, it can spill over from subject to object and object to subject, violating the bodily limits upon which classic subjectivity depends, and so smoothing the way for projection and introjection' (1988, 80).[8] Interiors (feminine) and exteriors (masculine) are confused by acoustic mirrors, making audible the gaps, fissures, and splits that are integral to subjectivity. The spatial coordinates of the gendered bodily ego are undone. This undoing is felt to be regressive. 'By returning to a moment prior to the entry into language and the articulation of subject/object relationships, it attempts to fuse mother with child and so to abolish the opposition of inside [feminine] and outside [masculine]' (ibid., 99–100).

Reverberations in the toilet are uncanny. There is nothing like a maze of stalls, metallic walls, glass mirrors, and urinals to confuse identity. The architecture invites auditory and visual confusion, as it also enforces what are for many arbitrary and illusory mappings of the body by acoustic design. Lavatory designs (much like Hollywood cinema) 'situate the female subject firmly on the side of spectacle, castration, and synchronization, while aligning her male counterpart with the gaze, the phallus, and what exceeds synchronization' (Silverman 1988, 50). Heterosexist and transphobic laws of symmetry and dissymmetry place great demands on trans and cissexual women alike to identify with femininity. Having to abide by an acoustic concord not of one's own making is disconcerting and constraining with respect to the accessibility of public space.

Voice was, for many interviewees, a requirement for access. Interviewees were often compelled to talk to gain entry to the feminized toilet. Female voices ratify feminine gender positions because they are, as Silverman (1988) suggests, associated with the vagina, interiority, and with 'organ holes' (67). The feminine 'voice provides the acoustic equivalent of an ejaculation, permitting the outpouring or externalization of what would otherwise remain hidden and unknowable' (ibid., 68). To be overheard by others is an intentional act. It is not a neutral or passive act. It is, for some interviewees, a way to parody a heterofeminine gender identity so as to gain access to the 'women's' toilet. Callum, who is a trans guy, says that he and his partner would 'just ... start talking about stupid random shit that we wouldn't be talking about otherwise ... because somebody has to hear my voice to know that it's okay that I'm going to be using that [female] bathroom.' Erik Prete, who is genderqueer and mixed Ojibway, is infantilized while imposed upon to speak:

> [People say to me] ... excuse me, like, little boy. Have you lost your mommy? And ... actually I'm twenty-three and I don't know where my mom is ... .you're trying to pee ... and ... people ... feel the need to ask ... you something so they can hear ... your voice. So ... they can decide what your gender is ... they're trying to think up something they can ask you.

Rohan recounts a frustrating experience:

> I got stopped by a janitor going into the bathroom ... one time, on a break from class [at university], and the janitor ... saw me walking down the hallway, and he was, like, 'That's the women's washroom,' and I just didn't have any patience that day, and I was, like, 'Yes, and I do have the appropriate genitals to be using this particular washroom. Would you like to see them?' And he was, like, 'Oh, sorry.' So I think as soon as he heard my voice he knew he had made a mistake.

For transmasculine, butch, and trans male interviewees, voice was instrumental to accessing the feminine space. But for many transsexual women, voice was a detriment to access because it did not chime in tune, pitch, and tone with cissexist assumptions about feminine acoustics. As Phoebe, who is a trans woman, remarks,

> There's some things I shouldn't do, like speak. I remember one time, not too many years ago, a bunch of girls came into the washroom, and they were giggling, and they did something really funny and I was laughing, and then [one of them said:] 'There's a man in here! There's a man in here!' I was so embarrassed. Like just my giggling, the way I laugh. There's an assumption of what a masculine or feminine voice sounds like in that interpretation of my giggle. But, I mean that didn't feel good.

Lana, who identifies as a non-trans genderqueer femme gimp, has a personal assistant accompany her to the toilet. She notices how speech gives her assistant's gender identity away:

> There's this one particular guy, who's like a very awkward straight guy ... he feels very uncomfortable going into the women's restrooms [with me] and ... he doesn't want to talk ... his voice is giving him away in that moment ... [I'm] like, 'Could you hand me that [toilet] paper?' And ... [then he is] ... not passing anymore ... if he talks then his presence is known ... there's a lot of talking and just like the logistical talking [with the attendant] ... like where the bowl is and what goes where ... those kinds of things ... negotiating ... the lift and there's also ... just chatting, very social endeavour for me, it always has been ... So, it's strange when you go to a public bathroom and then people feel uncomfortable to talk because then [their gender] will be discovered.

It is important to note that gender symmetries are read through racist imagery and acoustics. For example, Madison relates a troubling incident:

> My old adviser ... did ... [not] ... look ... identifiably female ... she was also Asian. So, people constantly thought that she couldn't speak English and that she was reading the labels on the bathroom [doors] wrong ... sometimes people would yell, because that's what you do when people can't speak English [she says sarcastically]. So, they'd be, like, [interviewee yells] 'Wrong Bathroom!' And I'm sure that it feels harassing at times ...

she definitely talked about [how she will now] ... go into bathrooms ... whistling so that people can hear ... [her] voice and identify ... [her] ... as female because if ... [she doesn't] ... people yell ... [about how she is in] ... the wrong space.

Trans interviewees of colour employ many different strategies to deal with gender-based questioning by white cissexuals. For example, Neil, a transmasculine interviewee of colour, often refuses to speak in toilets at all because his voice is heard to be at odds with his masculine gender identity (as I discussed in chapter 3). His refusal to speak troubles those who are looking for a cissexist gender symmetry in the domain of sight and sound. As Butler (2005) notes in her discussion of recognition in *Giving an Account of Oneself*, silence 'calls into question the legitimacy of the authority invoked by the question and the questioner or attempts to circumscribe a domain of autonomy that cannot or should not be intruded upon the questioner' (12). In other words, Neil refuses the acoustic call to give a normative (cissexist) account of himself. In his refusal to respond, he also challenges the legitimacy of the addressee and his or her demand to the other, who is, in this instance, transmasculine. It must also be said that silence gestures to a horizon beyond the visible; to a place of indeterminacy beyond signification. The performative non-speech confuses white cissexual body maps, which depend upon a normative concord between image and sound to apprehend gender identity, and it gestures to a mode of being that has not yet been normalized in the social field.

Masculinity and the Phantom Phallus in Space

In his discussion of body-ego spaces, Steve Pile (1996) reminds us that 'bodies occupy, produce themselves in, make and reproduce themselves in multiple real, imaginary and symbolic spaces, which are never innocent of power and resistance' (209). Speaking of hegemonic, white (non-trans) masculinities, he says that the bodily ego is continually projected into space, territorializing and usurping the space of the other, the feminine, and female sexuality. Arguing that object relations are projected onto public spaces, he suggests that a primal Oedipal fear of the mother is to be seen in a widespread preoccupation with feminine engulfment. Symbolic castration is offset by a hypermasculine display of the penis and a concurrent shunning of the feminine symbolized by open, ill-defined 'gaping bodies' (208). Heteronormative and cissexist masculinities can be understood as the product of a dual emotive and psychic composition involving 'a fear of becoming lost in unmapped and unmappable body-ego-spaces and a desire to penetrate and conquer those body-ego-spaces' (ibid., 209).

The politics of penetration and conquest, exteriority and interiority, masculinity and femininity are negotiated (sometimes violently) in the toilet. Perhaps the most poignant signifier of the gendered composition of the struggle is to be seen in the insistent production of and claim to the phantom phallus, a *symbol* of power to be distinguished from the penis as body part (which is flesh and mortal). The phallus is apprehended when one can use the urinal in an upright (straight forward) position and sound urination. As Alexander Kira (1966) notes in his discussion of public toilet design, the urinal resembles a vagina 'both in terms of the shape of the fixture and the nature of the act of urination' (55). The phallic production orchestrated by restroom design and choreography is thus heteronormative as it is also cissexist. To gain access to the 'men's' room, interviewees say they have to cultivate an imaginary penis and refuse body sounds that may be intercepted as 'feminine' or 'gay.' Auditory spatial significations impact upon the way genitals are apprehended. Erik Prete narrates a very common problem: 'With [trans] men who can't stand up to pee yet. Or ever. You know, they're gonna get caught. You know, someone hears them peeing.' As Isaac, who is a trans man in transition (and taking hormones), says, 'There's always a slight nervousness ... [I worry that] if I'm peeing in the stall, will someone think I don't have a penis ... maybe it sounds different!' Another female-to-male transgender interviewee concedes that 'I still feel uncomfortable because I sit to pee, and I just feel nervous when I'm going; nervous about whether I'm passing or not.' Seo Cwen, a non-operative trans woman (taking hormones), says that while using the 'women's' room, 'I used to worry about ... whether there might be a significant difference in the sound of pee trickling into the toilet water depending on which set of genitals it came out of and whether that difference might give me away.' Property, who is a non-trans queer femme, notes that some trans men squat over the toilet seat to make it sound as though they are standing to pee. She recalls that 'someone was ... [talking to] me recently ... about squatting in [the 'men's'] bathroom [stall] ... because if you squat over the toilet, it sounds like you're standing up to pee.' Rocky, who is genderqueer (and not trans), says, 'I have read in the trans pages – like, community-based pages – how a lot of FTMs are worried about going to the bathroom sitting down ... One guy even said I am going to be called a faggot. I was, like, that's interesting, because he's sitting down.' David even says that 'In an ideal world of toilets ... I would construct [a] ... proper acoustic environment ... preferably [with] ... white noise.'[9]

Worries about urinary acoustics are, of course, legitimate. Trans men and transmasculine people have to be concerned about physical safety in ways that cissexual men who are conventionally gendered, white, able-bodied, and heterosexual do not. Femininity is, as Silverman (1988) and Pile (1996) agree, about interiority, enclosure, and confinement in public space. The metaphor of the closet belongs to homosexuality as that of the bathroom stall belongs to the feminine. While trans and non-trans men alike use the stall to urinate for a variety of reasons (having to do with fears of urinating in public, not wanting to become a homoerotic spectacle for other men, wishing to avoid splash-back, etc.), use of the stall rather than the urinal is more likely to be read as feminine. Not being able to use the urinal can upset one's claim to masculinity. The refusal to be confined to an interior space is one reason why doors on stalls are often removed or broken in the 'men's' room.[10] For those whose rights to masculinity are already tentative and fragile in transphobic cultures, the lack of cubicle doors is an additional barrier to accessing public facilities.

If, as Pile (1996) argues, there is a psychoanalytic of space, a means by which the body is projected into space, then it is possible to consider how gendered bodily egos are mapped onto lavatories. Fred Rush (2009) suggests that there is a relationship between bodily or sensory rhythms and modern colours, lines, and spatial designs.

> So, one might rightly suggest that bodily rhythm is affected by change of perspectival line, parallax vision, the hue of color on walls, the feel of the building material, and the play of shadow and light at different times of day, in a way that is similar to the impact of melody, thematic alteration, tonality, and the 'color' of instrumentation in music. (Rush 2009, 100–1)

While considering the use of acoustics to map, avow, and delimit one's position in space, it is important to ask how the toilet may be structured by sonorous Oedipal grids. Early body memories of having been swaddled and intrauterine are projected onto space in ways that echo and reverberate. Female alignment with interiority is, according to Silverman (1988), a 'defensive reaction against the migratory potential of the voice – as an attempt to restrain it within established boundaries, and so to prevent its uncontrolled circulation' (84). Feminine vocal acoustics are a threat to hegemonic and typically cissexual masculinities because they stand in for the maternal voice. The 'maternal voice is also what first ruptures plenitude and introduces difference, at least within the paradigmatic Western family – the voice which first charts out and names the world for the infant subject, and which itself provides the first axis of Otherness' (ibid., 86). But it is also the extent to which the maternal voice orbits outside the infant's range of vision or capacity to see that is at issue, because auditory functions precede visual functions.

The cissexual man's proximity to the maternal, or to the feminine voice (as this designation takes the place of the mother in adult life), is architecturally regulated in bathrooms. Part of this architectural divide is governed by the acoustic containment of the feminine voice or body sound. There is often a refusal on the part of the cissexual masculine subject to be bound to the stall. This is because a cloistered noise designates interiority or invagination. But the feminine must also reside where she can be seen and heard in concert. The disordering of feminine voice and image conjures up foetal recollections of the maternal uterine wall that can be worrisome to the masculine subject (Silverman 1988). Part of what happens in bathrooms, in Hollywood film and in Western cultures in general, is a reversibility of the infantile or foetal scene: the mother (or feminine subject as stand-in) is placed in an 'enclosure like the one she herself provides ... locating the mother where the fetus in fact belongs – inside the womb' (Silverman 1988, 108). The containment can be understood as a masculine defence against maternal engulfment. The cloistering of the feminine voice and the externalization of the masculine voice (or body sound) are major North American cultural preoccupations. Given the pervasiveness of the defence, it is imperative to understand the injuries it can inflict. While not all cissexual heteromasculinities are consolidated by an obsessive need to distinguish the self from the feminine subject, there is an ever-present potential for violence in venues where the masculine territorial claim is weak. In such rooms there will be a heightened defensiveness that manifests itself in an obsessive interest in sexual difference – its identification and policing. There is a wish to repudiate all that is 'despised ... most abject, most culturally intolerable – as the forced representative of everything within male subjectivity which is incompatible with the phallic function, and which threatens to expose discursive mastery as an impossible ideal' (ibid., 86).

The Vaginal Mirage in the Acoustic Mirror

Feminine synchronicities depend upon what we might call a vaginal mirage or acoustic still. As argued above, female genitalia are hidden from view, their sounds stifled in the cloistered stall; and yet though unseen (hidden from view), they are all the while conspicuously present. The vagina is coded as insignificant (a 'little penis') in phallocentric economies. Using Derrida's concept of 'invagination' to describe the process through which feminine subjectivity is consolidated, or, as Silverman (1988) suggests, 'inserted,' the author of *The Acoustic Mirror* writes:

> Derrida's definition can be applied with great precision to ... operations which equate the female voice with diegetic, psychic, and corporeal interiority ... a recess or enclosure into which woman can be inserted. And since this recess is always linked by analogy to the image of female sexuality as a bottomless pit, these operations could also be said to fold that sexuality 'into' woman, to be one of the mechanisms whereby she comes to be identified with a dark continent. (70)

The vagina is a symbolic hole (like the mouth) and enables projections and insertions of various kinds. As a negative space – racialized through its Freudian pairing with the feminine as 'dark continent' – the vagina must not be seen or heard.[11] Its presence (or vacuum-like hole) interferes with masculine sovereignty and autonomy in public space. There is, as Pile (1996) and Silverman (1988) agree, a masculine territorialization of space, visual and acoustic alike. The vagina (as a primary signifier of the feminine) must be unheard and unseen. Its absence must be performed. Like all performances, audiences must ratify the act. In the 'women's' room, folks pretend not to hear what orbits into auditory range. People will perform an acoustic silencing. The urinary echo is heard and not heard at the same time. Audiences will, however, abide by the urinary non-hearing for certain actors and not others. A splatter that is too loud (indicating a standing position), a trickle that is too long (indicating something more than a full bladder), or a leak that may reveal a penile point of origin will upset listening audiences. Only certain tintinnabulations are validated or allowed to remain 'unheard.' Those with penile parts (regardless of their transfeminine gender identities) will be subject to auditory amplifications. When heard to violate heteronormative and cissexist laws of symmetry, trans women are not seen or heard to be 'women.' Security and police are sometimes called (as discussed in chapter 2) because a penis (or phallic part) must not occupy what is to be a negative vaginal space. (Of course, many trans women have vaginoplastic surgeries and thus 'no penis,' but the cissexual refusal to unhear a urinary stream puts one's access to the 'women's' room at risk.)

Urine splatters must be unheard in concert with the unseen vagina. The synchronicity performed involves a vaginal acoustic. There is to be an absolute temporal concordance between the urinary sounds emitted from the body and a visibly female gender identity. This performance is difficult for some trans women, who may not easily mirror tempos or acoustics that jibe with the sound protocols authorized by cissexuals in the feminized toilet. Lana explains how gender recognition by urinary acoustics can be a problem:

> People could make an assumption based on listening to how someone pees about what their gender is ... Occasionally, especially in women's bathrooms with the stalls, where sound is so much more [an issue] if you can hear someone in the next room using the bathroom ... [you're] probably making assumptions about, like, womanliness, right?

Gypsey notes the acoustic implications of urinary positioning:

> If the person is standing up to pee ... it's a different sound. And you know [people decide] if it's a guy or a girl in there [on the basis of noise] ... the girl next door [may] think you are a guy because it [urine] makes a different sound. It's got longer to travel and it sounds different.

Temperance, who is a non-trans queer femme, confesses that while in the cubicle, 'I listen to the sound of ... peeing [next door] to determine how far up from the seat they are sitting.'

Interviewees also sought to dissociate themselves from menstrual noise. Socially coded as pollution, menstrual blood evokes fear, loathing, and disgust. Menstrual blood is both a signifier of disorder (inner body fluid appearing outside the body) and a symbol of mortality (death and dying), and it also conjures up images of the maternal-feminine (as omnipotent figure, wielding her powers of life and death). Anthropologists have also argued that taboos on menstruation reflect fears of the feminine, of its capacity to usurp or impinge on masculine space (Gillison 1980; Meigs 1984). If normative femininity is measured by a refusal to take up or to impinge upon masculine space, then menstruation will be coded as 'dirty,' something to absorb, hide, and sanitize.

Abject menstrual sounds are subject to great consternation. However vigilant women may be about personal hygiene in toilets, they are likely to be heard or seen to be dirty. Alignments with dirt call feminine gender identifications into question. Menses seen or heard interfere with subject integrity and code the feminine body as dirty (matter out of place). Temperance emphasizes the noise made by 'sanitary' products:

> Opening a pad, like a commercial ... maxi-pad makes a noise, makes a plastic rip. It's audible ... everyone knows that you are opening a pad, and yet that's dirty. That act is dirty. You cover up the sound, you muffle it. I used to rip my pads open and then close them so that I wouldn't have to make that sound when I put on the pad. I think that's kind of ... abject, menstrual products and menstrual waste.

As another non-trans queer femme interviewee recalls, 'Like, if you're caught ... overheard unwrapping a tampon [in high school] that was really perilous.'

Female emissions interfere with gender and subject integrity more so than male emissions in public toilets. Faecal plops, the sound of urinary splatters or of sanitary pads tearing, all conjure up images of the maternal feminine. Silverman reminds us that, for Lacan, the maternal voice, 'the feces, the mother's breast, and the mother's gaze, designate objects which are first to be distinguished from the subject's own self, and whose "otherness" is never very strongly marked' (1988, 85). These objects (what Lacan refers to as *object (a)*) are felt to be parts of the self. As they are lost through maturation and subject individuation ('growing up' and separating from the mother), they are psychically retained as 'lost' or 'missing body parts' (ibid.). 'Once gone, it [*object (a)*] comes to represent what can alone make good the subject's lack' (ibid.). That which is lost to the subject is either idealized and apprehended to fill a lack or gap at the heart of the human subject, or rendered abject. The female voice can, for example, ignite desire (a quest for a perfect unison) or be heard as a maternal horror threatening to engulf the masculine subject. There is a fine line between the ideal and the de-idealized feminine voice or part object.

As Kristeva tells us, 'abjection is elaborated through a failure to recognize its kin; nothing is familiar, not even the shadow of a memory' (1982, 5). The abject is that which is repudiated but also that which can never be fully divorced from the self.[12] Abjection and disgust are both structured by forfeitures of memory, and are consequently dissociative. Similarly, in his cultural analysis of disgust, William Miller (1997) notices that disgust, like abjection, is not *just* about 'being too close' to something culturally vile, for example, but also about a disowned memory linked to a wish or a barrier to desire, or, alternatively, to a translation of desire *into* disgust.[13]

In the following two chapters, I write more about how bathroom architectures, designs, and choreographies manage dialectics of desire (affinity) and disgust (refused affinities). For now, it suffices to say that toilet designs protect typically cissexual heteromasculine subjects from Oedipal fears of maternal engulfment. Modern architectural designs, according to Steven Pile (1996), repeat separations from the feminine (coded as maternal), along with symbiotic interconnections between people. 'Cold abstraction produces a spatiality without pleasure, a castrated space – real and imaginary, concrete and symbolic' (163). Quoting Lefebvre, Pile pinpoints how 'Our abstract space reigns phallic solitude and the self-destruction of desire' (Lefebvre 1998, 309). The erotic is devalued along with the feminine because 'woman' is the sex that is ill-defined (or, to quote Irigaray, 'not one') (1985). Metal walls, locked doors, smooth, hard, sanitized surfaces – absolute spatial delineations concretized by acoustic designs and glass mirrors – all map the coordinates of gender in space. Much is done to offset the fleshy, womblike, or choric composition of the room so as to isolate and objectify a cissexist (and phallocentric) concord between gender and genitals.

Gender and Urinary Positions

Posture is apprehended by sound vibrations. Bathroom acoustics impose upon people to assume dimorphic 'urinary positions.' Gender is disciplined by architectural designs that mandate gender-specific postures and positions (not unlike missionary sex positions in the bedroom). It is worth noting that the urinary position one assumes in the lavatory is historically and culturally variable;[14] it only *appears* to be determined by genitals. Excretion is a performative ritual that consolidates an illusion or a truth about sex. Variations on the dimorphic poses are detected by sight and sound. Urinary postures are too often presumed to be outside the realm of societal construction and gender performativity (Butler 1990). As with traditional prohibitions on sex, bathroom choreography is gendered, authenticating some positions but not others: genders are to be oppositional (clearly delineated by posture), though complementary. 'Men' are prompted to stand upright and approach the urinal from a forward position; 'women' are prompted to sit, squat, and approach the toilet bowl from a backward position. This is an acoustic dance of mirrors authenticating heteronormative urinary positions in concert with dominant imaginings of heterosex. Fluids must flow in a particular direction (from male to female), from a particular orifice (the penis), into the vagina (or a white, oval toilet bowl gendered as a feminine receptacle), from subjects in gender-appropriate and heteronormative positions (over, up, and above the urinal or in a relatively confined sitting/squatting position). Urination is ritualized to the point where some people cannot do it in another position. Excretion is subject to a fixed choreography in time and pace with modern toiletry habitus.[15] Women are imposed upon to touch the toilet bowl, while men do not have to touch the urinal. Women tend to feel vulnerable in the stall, as men are expected to assume a potent and aggressive stance before the urinal. It is not a coincidence that 'interiority is also identified with discursive impotence, and exteriority (at least by implication) with discursive potency' (Silverman 1988, 75).

Illusions about genitalia are created by urinary positions and gender-specific choreographies. As one transmasculine interviewee notes,

> Women and men can both pee in different ways ... Like, women can certainly stand up, and men can certainly sit down. So it's interesting that there's these two [positions], and of course it's enforced by public spaces, like the way that women's rooms are laid out and the way that men's rooms are laid out ... they encourage ... [dimorphic] ways of urinating ... people are capable of both but encouraged to do one thing or the other ... I think it's just kind of reinforcing different [gender] practices ... [to] ... keep ... a rigid binary.

Another trans man notes how gender positions in the toilet are taught: 'But when men are growing up they pee sitting down at home, right? Because it's usually mom who is doing the toilet training ... right? ... it's easier. So I think that boys do learn to pee sitting down, but they pee standing up [as adults], I think, it's more of a social thing.'

While there are gender norms governing how one urinates, there are numerous ways in which the regulations are breached. According to Tara, who is white, transgender, and gay,

> I actually happen to know quite a few biological men that pee sitting down. And I actually know quite a few that will not use urinals, because they're too shy ... or they don't like peeing in front of other people. I know some that pee sitting down ... [Sometimes guys] don't wanna stand up and pee in front of the whole world.

Lana recalls that she was surprised to see men urinating from a sitting position: 'I've had my own moments where I've seen bio-men sitting down to pee and being like, oh! "I didn't know bio-men could do that!" ' Some interviewees say they are interested in how the so-called opposite sex pees. In the words of one bisexual interviewee, 'I am fascinated by what people do in bathrooms, behind closed doors ... you know, like how they urinate.' Many are curious about what happens 'behind closed doors,' intrigued by what people do in bathrooms, knowing full well that they can be sexualized spaces.

When one is recognized to be transgender, this curiosity is piqued. Eric Prete, for example, notices how 'people are ... amused [by excretion] ... Oh, how does a trans person pee? You know, like, to me that's ... childish curiosity [very much about the "discovery" of sexual difference].' What we do in toilets signals gender, even as we are unclear about what precisely is being signalled by any one urinary position. Emily confirms that trans people are attuned to how their bodies are seen and heard in toilets: 'Many of my trans friends who are ... [transitioning or transitioned] spend a good deal of time ... coming up with ways ... they can use urinals ... it's ... important ... [and it's] one of those important cues ... if ... [they] only ever used a stall ... you could get noticed.'

Using the urinal (as opposed to the stall) authenticates and validates gender in ways that are deeply personal. Sometimes the poses are accompanied by ambivalence. For example, as Gypsey notes, 'It's more natural [for me] to sit down. I am more comfortable sitting down than having to pee into ... [a urinal] ... Even in the male world when I lived in it, I did not prefer to use the urinal in a male washroom. I would use the stall and sit.' A substantial number of interviewees who use the urinal are uncomfortable with it because it ratifies a dominant heteronormative masculine position. As Chloe, a non-trans queer femme, notes, 'There's lots of rules about how people go about ... [using the urinal] ... I think it's expressing masculinity ... the way one approaches the whole urinal experience ... can signify whether one is straight or gay, or whether one wants to pass as straight.' Sugar, who is a cissexual woman, emphasizes how standing before the urinal is an essential feature of masculine heterosexuality: 'I think a lot of men feel emasculated when they pee sitting down.' She also says, 'You're more vulnerable if you're sitting down versus if you're standing up ... towering over somebody ... men in the urinal, they're ... not standing passively. They've got their legs wide apart, their feet shoulder width apart. It's a very dominant posture ... It's not passive.' A gay male interviewee puts it succinctly: 'If you just whip it out and pee, you're a guy.'

Gay male interviewees who are not trans often express reluctance to approximate heterosexist ideals of masculinity at the urinal. Some who cannot use the urinal, or feel uncomfortable doing so, sometimes express feelings of shame. Often they feel their masculine gender identifications to be at risk. As s. applebutters, who uses the 'men's' room, confides, 'I don't want to piss in front of anybody ... I guess there is more shame, you know?' One queer male interviewee underscores how difficult it can be to urinate in the presence of potential onlookers:

> If a urinal is well-constructed, you are not terribly exposed. But in some cases you'd have to wonder what people were thinking ... [because] these urinals are installed lower down and they don't have sides that come up higher and stuff. And they are positioned next to each other so you are pissing shoulder to shoulder next to these little basin things and that can sort of be uncomfortable ... men feel sort of pressured to ... urinate really quickly because other people are waiting ... paradoxically ... that can stop the flow of urine and [that can amplify] ... the difficulty of getting started ... .[I] question [whether] ... the man next to me notice[s] that it's ... [taking] me a long time to get started ... panic sets in [when] I can't pee at all ... [and so I] pretend [to have] peed and flush the urinal [and leave the room].

As another gay male interviewee who is cissexual says, 'Lots of guys, very confident guys, have a serious problem [peeing in urinals].'

Trans men also tend to be reluctant to use the urinal. As noted in the previous chapter, some transmasculine people are unable to use the receptacle to urinate. As a result, they use the stall and worry about how their gender will be apprehended. Callum elaborates: 'A lot of friends [who are trans] ... believe that if they don't use the urinal they are suspect.' Chloe talks about a trans friend of hers who is 'just really frightened ... [because] sometimes ... he doesn't use the urinal, and therefore ... [might] have a problem ... men do notice.' The 'men's' toilet is often inaccessible to transgender folk because urinals are difficult to use.

Consequently, trans interviewees frequently 'hold it,' containing or delaying urination for prolonged periods of time. As one trans two-spirited interviewee recalls, 'I started doing this [holding it] at such a young age I've probably trained my body ... I reached a certain age and then I started ... training my body.' Some interviewees do not use public restrooms at all and discipline their bodies not to need it. Sarah, for example, says, 'I always go [to the bathroom] before I leave [the house], and I become the champ in holding it for, like, a few hours.' One trans male interviewee similarly notes, 'I would time my day, or end up holding it for a long period of time.' Gypsey says she regulates her water intake: 'I have trained myself that before I leave home I use the washroom. And if I am going to drink a lot of liquids I am going to do it at home.' Velvet Steel, a trans woman taking hormones, reveals that she has 'actually peed my pants. Because I've refused to use the washroom just for whatever situation was taking place, or whatever reason I felt uncomfortable using it and tried to hold on as long as I possibly could. And, you know, didn't make it.'

That interviewees discipline their bodies by delaying urination is confirmed time and time again by the interview data. Toilets are often inaccessible to trans people because heteronormative and cissexist architectural designs and choreographies are restrictive. Because people can't always perform the synchronizations necessary to ratify gender (and others do not want to), some must avoid public facilities entirely. Emily, a trans woman, explains:

> Because of my neurosis about going into public bathrooms I've learned a certain amount of discipline ... I always make sure I go before I [leave my home] ... and I always make sure I'm ...hold[ing] off as long as I can until I can get home again ... I tend to avoid anything that might give anybody a reason to think I was male.

Not only does urinary discipline curtail or restrict one's movement in public space, it can also compromise one's health. Jacq, who is genderqueer and butch, believes that her kidneys may be compromised:

> Since I've gotten into my forties, I can't hold [it] as long ... Whereas before in my twenties and thirties, I was fine at holding it. I can't do it now. I'm good for about two hours and that's it ... I'm not really sure if I *do* have a problem with my kidneys ... But, uh, I probably do, from holding it.

Temperance similarly laments,

> It's tragic that people can't satisfy one of the most basic needs ... [toilets] put their health at risk ... their physical safety at risk ... [ironically they are] places meant to make your body healthier, in terms of keeping the routines of your body at work ... [but] those places ... [are] sites of violence [and exclusion] for trans people.

The necessity to delay or avoid urination is a damning commentary on the obsessive gendering of lavatory space by acoustic and mirrorical design. These designs interfere with the plumb line of the body. We need to plumb the body (eject fluids) to live. Plumbing *is* a matter of health and safety, while heteronormative and cissexist orderings of space are not. By imposing a life-denying, inessential rule of gender symmetry, we establish a transphobic zone of exclusion: a horizon or vanishing point designating people 'out of bounds.' The horizon is a limit at the cusp (or outer reaches) of visual and acoustic radars. The 'fundamental indeterminacy of self and other in perception' (Vasseleu 1998, 75), the differences between the masculine and the feminine, need not be architecturally regulated and subject to policing. The gendered subject is transitive. Let us enjoy and encourage what may appear to be an acoustic and visual discord or an unpredictable symphony. By striking new and different chords, we make more living space for those whose lives exceed the normative visual and acoustic concord. In the following chapter, I consider how the hygienic imagination functions to purify and delineate gender by tactile and olfactory mirror circuits.

5 Touching Gender: Abjection and the Hygienic Imagination

> Society scares easily at those aspects of sensuality that it qualifies as obscene ... *Inter faeces et urinam nascimus* (we are born between excrement and urine).
>
> (Bernard Tschumi, quoted in Lahiji and Friedman 1997, 36)

The panoptic designs of the modern toilet owe much to the plague. Cholera, diarrhoea, smallpox, and typhus were, as mentioned in chapter 1, big worries to sanitation reformers and city planners concerned about mortality rates in Britain and Europe. When it was known through scientific testing that sewage-contaminated drinking water led to disease, efforts were made to improve city sewers and to eradicate cesspools and faecal and urinary deposits in urban streets. The ultimate fears of the early-modern era were those of disease, contagion, and death – all of which were managed by order, quarantine, and partition.

Gendered toilet designs of today are rooted in the ways Londoners and Parisians managed disease, what Foucault calls the 'great confinement' (1979, 198). But the plague was not just a physical ailment. It was a rationale upon which people could be internally divided and subject to surveillance. 'Underlying disciplinary projects the image of the plague stands for all forms of confusion and disorder; just as the image of the leper, cut off from all human contact, underlies projects of exclusion' (ibid., 199). Worries about contamination were projected onto the body of the leper, the criminal (often thought to hide out in the underground), the prostitute (symbolically aligned with raw sewage, disease, and contaminating fluids), the destitute (who searches for sellable items buried in septic sludge), the vagrant (who slept in the city sewers), the scourer (who cleans city drains and sewers), and those racialized as degenerate. The trouble with disease, however, is that it does not discriminate. During the nineteenth century, epidemics of cholera and typhoid affected the bourgeois classes, the royals, and the well-to-do: 'Death seemed unwilling to bless the squire and his relations, and keep us in our proper stations' (Wright 1960, 210).

While there is no basis for gender-exclusionary designs in epidemiology, segmentation by sex in bathrooms today is often rationalized by recourse to ideas about public health and safety. Gender-segregated designs are sacrosanct because many people are preoccupied by the careful delineation of sex. This is not because one gender can infect the other but because gender disorder is sometimes *felt* to be a matter of life and death. Gender incoherence, or, rather, what is taken to be an incongruence between gender identity, the sex of the body, and the insignia on the bathroom door, is metonymically associated with disease. Sodomy is also associated with disease, HIV and AIDs in particular.[1] There is never enough soap and disinfectant to kill whatever it is people are afraid of catching. We are subject to quarantine and compelled to purify our bodies (literally and symbolically). We cleanse the boundaries between the masculine and feminine (or separate the two) in public lavatories so as to police the borderland or indeterminate space between these two discursive and material positions. Gender purity is disciplined by hygienic imaginations and rendered sacred, while gender impurity – signified by a discord between gender identity and the way the sex of the body is intercepted by others – is profane. 'Modern cleanliness departs from ancient ablution in its extension of hygiene to the psychological interior' (Lahiji and Friedman 1997, 42). Hygiene is no longer a ritual or set of practices exclusively focused upon the material body but a pedagogy or art of government targeting gender in ways that are psychically significant.

In their discussion of modern architecture, the sink, and abjection, Lahiji and Friedman (1997) coin the term 'hygienic superego' (11) to illustrate how cleanliness is tied to the law, and to prohibitions on sensorial pleasures (other than vision). The hygienic superego polices the gap between purity and abject dirt (defilement) (ibid.). What is pure and abject is no longer (or, rather, not exclusively) determined by hygiene; it is about gender coherence. 'Prohibition against dust and dirt marks the structure of the hygienic superego. This prohibition is aggressive; it propels modernism and identifies with it. The clean body is also a plumbed body' (ibid., 42). But the plumbed body is also a carefully sculpted and coherent gendered body; one that cannot be confused with the 'other' sex. Because gender is about how we seal and delimit the body, how we navigate identifications and desires in relation to others, it should not be surprising that hygienic superegos are focused upon governing orifices and genital zones which are points of interconnectivity. Panic about gender and panic about sexuality intersect. Injunctions placed on homo, queer, or perverse sex demand that olfactory[2] and tactile sensations be stifled – noses plugged and fluids kept at bay. When abject body fluids commingle, gender is sometimes felt to be at risk. Those who are seen to be impure – specifically those who are recognizably LGBTI – are sometimes perceived by heteronormative and cissexual folk as contaminating the public body because they allegedly ignite otherwise dormant sensory registers. Abject desires, those regarded as repulsive in the bathroom – such as hygrophilia (pleasure sought by physical contact with body fluids) and mysophilia (arousal by the inhalation of body secretions) – are so constituted because of the now widespread degradation of sensorial pleasure (touch and smell in particular). Erotic pleasures that involve body parts, orifices, scents, and fluids that do not abide by heteronormative and cissexist prohibitions on desire are disowned or, literally, driven underground.

Tactile and olfactory sensory systems are subordinated (rendered impure) by modern optical and acoustic designs. Modern optics and acoustics are accentuated in toilet designs because, unlike other sensory systems, they enable distance and objectification. As Laura Marks (2000) suggests in her study of film and embodiment, modern optics is a less intimate or sensuous kinaesthetic than touch or taste or smell. Hearing also preserves objectivity and is mediated by air, as vision is mediated by light. For the auditory canal, the 'exteriority of its object is preserved even as sound enters the ears' labyrinths, because the sound in itself conveys nothing but the meaning given it' (Vasseleu 1998, 100). Those at odds with the cissexist and heteronormative body politics that mandates gender purity and coherence are held up to a bright, investigatory air or light in the lavatory, much like an amoeba under an open-air microscope.

The Gender of Abject Fluids

> The abject, like the uncanny, offers a valuable means to demonstrate the connections between psyche, body and society, and the way in which these are sustained spatially, both at the level of the individual and within

the surrounding social system. Boundaries, borders, and the very design of the social environment symbolize the fragile division which sustains identity. (Wilton 1998, 179-80)

Woman, toilet: these are the apparatus by which we are undone and which we abjure, in order to be who we are. (Morgan 2002, 175)

Gender purity is established by abjection. Julia Kristeva (1982) defines the abject as that which 'does not respect borders, positions, rules. The in-between, the ambiguous, the composite' (4). The abject opposes the 'I' and exists in a 'place where meaning collapses' (2). It threatens the modern subject at its constitutive boundaries. By abjection, we rid ourselves of dirt and substances that are impure or unclean. The abject (or that which is defiled) is also that which is 'jettisoned from the "*symbolic system.*" It is what escapes that social rationality, that logical order on which a social aggregate is based, which then becomes differentiated from a temporary agglomeration of individuals and, in short, constitutes *a classification system* or *structure*' (ibid., 65). Building upon Kristeva's analytic of abjection, Butler (1990, 1993), McClintock (1995), and Thomas (2008) further elaborate upon what they call social abjection to understand how people devalued in modern, Western industrial and capitalist nations are metonymically associated with abject body fluids, or, to be precise, treated 'like shit.' People are excrementalized. Late-modern societies expel and excommunicate people deemed to be unclean. The social body, like the individual body, polices its borders.

While there is no one-to-one correspondence between what we abject (and find grotesque) and desire in the realm of object relations and gender identity (trans or cissexual), there is, as interviewees note, a way that gender is secured by abjection. Gender is partially ratified by what (and whom) we abject and see to be other or different from the self. By aggressive disidentifications (you are nothing like me, bear no trace of or relation to me) or projective identification (whereby a subject projects unwanted parts of the self onto others), people police gender identity. There are gender-specific choreographies modelled upon what Inglis (2002) refers to as 'toiletry habitus.' These choreographies are evident in bathroom designs and in one's orientation to base body matter. It is frequently the case that one gender is thought to be more 'dirty' or 'unclean' than another, and the cleanliness of the bathroom mirrors these assumptions. Tara, who is a genderqueer butch, notes that 'women's' toilets are always cleaner than 'men's' toilets, and asks, 'How do men's bathrooms get so dirty and filthy?' A bisexual interviewee hypothesizes that 'women are more likely to complain about dirty bathrooms, and ... men that complain about dirty bathrooms are going to be seen as "effet" [effeminate] ... or feminine.' Sarah, who is transgenderist, speculates that we have gender-segregated toilets because 'guys are so dirty, guys need their own bathroom ... let them be dirty, we don't want to be affected by that [dirt].' As one trans man notes, bathroom designs cultivate illusions about the 'dirt' and 'stink' of gender:

> Just from the condition of bathrooms [it appears that] women's bodies are considered hygienic or made hygienic and men are considered dirty or stinky and that's okay ... you know, never a paper towel to be had and no soap [in the 'men's' toilet] because you don't need it you can just stay smelly ... it's fine. But in the women's room there is everything you need to [keep clean].

Some trans and non-trans women intimate that hand washing (whether or not one does it in the 'men's' room) is largely determined by the design of the porcelain receptacles (and the presence of toilet paper). In other words, the pedagogy or art of hygiene built into the lavatory is gender specific. As Emily, who is intersex, says,

> There's a very good reason for ... why men tend not to wash their hands nearly as much as women do. There's no toilet paper at the urinals, for starters. You don't really need it ... and so men don't tend to think of ... [their hands] in contact with anything they would think of as dirty ... whereas going to the bathroom as a woman, you are definitely going to get your hands wet ... it's a much messier experience. And ... there is much more of a visceral ... drive for women to get their hands washed ... so ... women are much more sensitive and aware ... of less clean bathrooms.

As interviewees note, personal hygiene is gendered and mediated by toilet designs. The difference between male and female, masculine and feminine, is authenticated by recourse to cleanliness; the 'ladies' room is imagined to be clean whereas the 'gentlemen's' room is thought to be unclean.[3] Because women are more likely to be read as 'dirty' and 'polluting' in Western cultural folklore (Grosz 1994, Kristeva 1982, Longhurst 2001, Shildrick 1997) than men (unless, of course, those men are racialized or classed as 'dirty'),[4] it is likely that women in general are held to more exacting cleaning rituals than men.

Interviewees also have much to say about colour and its relation to hygiene. They note that toilets are either painted or lacquered white to denote cleanliness, or are in muted pastel shades, normally pink or blue, to signify gender. Images of gender purity (cleanliness) tend to be denoted by white or pastel colouring. Butch Coriander, who is a non-trans genderqueer butch woman, explains that public toilets are 'white broken up by a colour of some sort, some sort of pastel ... like blue, maybe a light pink ... It's supposed to be pure and clean. White is supposed to be pure and clean.' Tara notes that 'White has a psychological ... association with cleanliness ... And ... it looks dirty faster ... When it's white, you can see dirt. When it's black, you can't.' Toilets kept white by elaborate bleaching and disinfecting rituals are said to be racialized. Sugar, who is a non-trans queer femme, surmizes that

> Cream and beige ... I think part of it's racial. Part of it goes back to that sort of 1940s, 50s, 60s, white is clean and good and we will do all in our power to bleach and whiten everything ... whether that's people or our houses and washrooms and ... I think that has informed paint colours in washrooms ... I think it's largely informed by ... needing to be white and clean and sterile and ... normal.

This normality is colonial and puritan in its emphasis on virginal and pristine toilet space. As an interviewee who is bisexual notes, 'Everything [in the lavatory] is always white ... which is really impossible to keep clean, and that's the whole point, it has to be pristine ... I think the ideal [sought] ... is that every time you go in [to the bathroom] it should look like no one has ever used it before.'

Whiteness signifies absence, or perhaps a vanishing point or horizon beyond which nothing can be seen. It is, as Richard Dyer (1997) writes, associated with 'purity, cleanliness, virginity, in short, absence' (70). Whiteness may also signify a dead end[5] or sensory limit. Colourful designs and ornamentation were characteristic of early Victorian pedestal closets and public urinals (such as the 'Gents' at Sough End Green in Hampstead, built in 1897, near the London and North Western Railway and used to stage public sex scenes in the film *Prick Up Your Ears* based on the life of the late, gay playwright Joe Orton). But these lavish designs disappeared as a capitalist ethic of utility, time management, and efficiency took hold. Loitering at the urinal and taking one's time on the potty were discouraged. Homosexual sex was subject to censure. The public lavatory was not to be a place of erotic contemplation, sensuality, or relaxation. Colour and lavish design came to signify unsavoury pleasures and were, over time, seen to be incompatible with public hygiene and prohibitions on public sex (sodomy in particular). Straight lines, metallic walls, and plainly tiled white surfaces replaced lavish Victorian water-closet designs, their circular patterning, ornamentation, and 'decadent' colour schemes.[6]

The visual contrast between the clean, white bathroom and the defecating body was amplified. 'Whiteness as an ideal can never be attained, not only because white skin can never be hue white, but because ideally white is absence: to be really, absolutely white is to be nothing' (Richard Dyer 1997, 78). This is as true of human skin as it is of the lavatory designs and fixtures inspired by the present-day hygienic imagination. Spectacles of death and transcendence (encapsulated by the story of Jesus Christ), flesh and spirit, darkness and light, impurity and purity, all haunt Western Christian nations and turn up in the way city planners, capitalists, architects, and engineers employed by large corporations build, design, and dictate how the populace will use public washrooms. As Joel Kovel notes,

> the central symbol of dirt throughout the world is faeces, known by that profane word with which the emotion of disgust is expressed: shit ... when contrasted with the light colour of the body of the Caucasian person, the dark colour of faeces reinforces, from the infancy of the individual in the culture of the West, the connotation of blackness with badness. (Quoted in Richard Dyer 1997, 76)

That the ethic of gender purity is colonial is evident in the historical example of Pears' Soap, which Anne McClintock (1995) uses to illustrate 'commodity racism' in Britain; in present-day hygienic rituals that remove not only bacterial build-up in public restrooms but people culturally coded as 'dirty' and 'unclean' (often the under-housed and street active – it is no coincidence that the economically dispossessed are called 'bums');[7] and in the criminalization of sodomy (homosexuality was imagined to be a eugenic defect, and consequently a danger to what Lee Edelman [2004] calls 'reproductive futurity,' impinging upon the general health and well-being of the population).[8] Gender purity is set up against the trans and/or queer subject as whiteness (instrumental to gender purity) is set up against those racialized as non-white and impoverished (under-housed and unemployed). Those without employment and access to affordable housing, and thus dependent upon public facilities to clean themselves and to get drinking water, are sometimes branded unproductive, 'faecal' parasites.[9]

Public facilities separate the body from its faecal remains; but they also separate the so-called upstanding citizen from those culturally coded as abject. In her reworking of the Kristevian notion of abjection, Anne McClintock (1995) argues that modernity produces abject 'objects' (like the anus), abject 'states' (like coprophilia), abject 'zones' (like the toilet), and socially designated 'agents' of abjection (like LGBTI folks, those with disabilities, those who are street active, those who are racialized as non-white, etc.) (72). 'Under imperialism ... certain groups are expelled and obliged to inhabit the impossible edges of modernity: the slum, the ghetto, the garret, the brothel, the convent, the colonial bantustan ... [and I would add the sewer, the urinal, the common latrine or cesspool]. Abject peoples are those whom industrial imperialism rejects but cannot do without' (ibid.). The crucial point to be made is that people are symbolically coded as abject (not just substances)[10] and abjected (sometimes literally from public space). In the case of the toilet, non-trans folk sometimes impose upon those who are perceived to be trans and/or queer to internalize, or to forge an identification with, that which is abject or culturally de-idealizing. In this way, cissexuals who are not gay-positive or trans-positive transfer their own gender identificatory troubles (and refused desires) onto trans and/or queer people. LGBTI people become the 'untouchables' of the toilet.

Prohibitions placed on touch and smell are about the management of ego boundaries. In her seminal notes on projective identifications, Melanie Klein notes that the one who projects unwanted elements of the self onto others may experience a 'weakened sense of self and identity' (Hinshelwood 1989, 179), and that aggressive disidentifications with the other are spurred on by envy. Distance taken between people may indicate not only objectification and distancing but subject-object confusion. Worries about whiteness and sanitation are, as I suggest in what follows, driven by anxiety about gender coherence (its purity and legibility). The degradation of touch and smell (the more intimate senses) is accomplished in part by the valorization of vision and, to a lesser extent, hearing. The optical design of the toilet is meant to patrol the distance between self and other. The bathroom places occupants in a fluorescent spotlight 'so that they are clearly separated from their surroundings' (Richard Dyer 1997, 86), from abject body fluids and people coded as abject in the normative landscape.

In his discussion of the modern bathtub, William Braham (1997) notices that rules of hygiene are not just about health and safety but about visual integrity: one must be able to see the body unencumbered by dirt. 'The appearance of the modern [bathroom] surface – smooth, white, shiny, sanitized – offers sufficient guarantee of protection from disease' (217). But the glow and appeal of the oval toilet bowl receptacle, urinal, or sink basin are illusory, offering only an imaginary defence against subjective entanglement (or exposure to others). Smooth, white, porcelain tubs, toilet bowls, and urinals are desirable because they symbolize, mirror, and refract a neutral tertiary space where the body will, presumably, not be exposed to the mess and spillage of others. We pretend that the underlying worry about touching the toilet bowl, for example, is about personal hygiene, all the while forgetting that urine is a sterile substance. The rim of the toilet bowl must *appear* to be clean. This is not because people worry about disease and infection (although we do), but because people are anxious about whiteness (denoting purity) and gender integrity. When abject body fluids are left behind by others, one's own gender integrity is sometimes felt to be compromised. In other instances, one may come to question the sanitary practices of others when the rules of hygiene are violated. This questioning is gender specific. Phoebe, who is a trans woman, explains that in the 'women's' room,

> It seems to me that this is a fairly constant behaviour of leaving toilet paper on the floor, tampons, like not throwing them in the disposal properly ... I'm just totally surprised because this ... is not the public demonstration of femininity. Femininity in its public form is considerably different. And so, I was surprised once I entered that ['women's' room] space, surprised, hell, I was shocked!

Phoebe underscores how feminine gender performances are dependent upon elaborate hygienic rituals and how these performances are interrupted by leaving abject fluids behind (or backstage, in the stall).[11]

Hygiene practices are moored by our openness to what I call cultural infection: that is, the fact that our bodies can be seen to be carriers, signifiers, or agents of abjection. While there is no one universal substance that offends in all moments (faeces notwithstanding), heterosexual matrices prescribe and set parameters upon the grotesque and the sublime. These parameters are governed by prohibitions and taboos relating to excretion and excreta.[12] Anthropological data confirm that there are widespread beliefs about how 'each sex is a danger to the other through contact with sexual fluids' (Douglas 1966, 37). The management of elimination tells us much about the hygienic superego, its gender and psychic structures.

Normative performances of disgust are gender, class, and racially specific. They reveal the inner workings of white hygienic imaginations. Disgust with urine is often about a perceived encroachment upon the border between inside and outside, private and public, self and other, masculine and feminine, white and non-white. This encroachment is evident in interviewee comments about urinary and faecal remains in the toilet. As Tulip, who is a non-trans genderqueer femme, says, 'I feel like women's bathrooms are dirtier and my theory is that women squat ... I would never sit on a ... public toilet ever ... there is more urine on the seat in women's bathrooms than men's.' KJ, who is a trans man, complains that '[Men] ... miss the friggin' toilet bowl ... So it's on the walls ... It's horrific! It's, like, jeez ... how did you do that? ... It's notorious – it's ... all guys ... the men's washroom is going to be gross, and the women's is going to be not gross.'[13] As one trans guy laments, '[In] some of the [men's] bathrooms it looks like they're [crap testing] when they're peeing ... it's all over ... the stall ... why can't [they] all just pee in a bowl?' Tara agrees: 'They just whip it out and whiz anywhere they damn well please ... it's pretty gross.'

One of the reasons urinary spill and splatter are a point of contention in restrooms is because they are not obviously gendered. The body's interiors are curiously ungendered (with the exception of the reproductive organs, differing chromosome counts, and estrogen and testosterone levels). Despite beliefs that urine's smell, colour, and consistency vary by gender, it is not a fluid that can easily be traced back to a given subject. Everyone shits and, like shit,[14] urine is a great equalizer.

Unease about urine is often played out upon the toilet seat, and obsessive attention is paid to the vertical (upright) or horizontal (downward) position of the lid. The way a toilet seat is left in a private bathroom signals the urinary position assumed by the previous user. Obsessive worries (usually in-home) about the lid of the toilet are commonplace and sometimes comical.[15] Consider the following memory of Tulip after immigrating to the United States:

> At home there were no rules or anything [about 'urinary positions'] ... when we got to the States from Israel ... my mom would always comment that the TV shows always had the wife and the husband fighting about

leaving the toilet seat up, which was really funny because we were a family of three women and one man and my dad always sat and the one time that he would stand up he would always leave the toilet seat up and nobody cared, you just put the toilet seat down. And then we came to the States, and it was this constant joke on sitcoms and, like, 'Oh, I am going to divorce you if you don't put the seat down.'

The state of the porcelain receptacle after use may be seen as comical, or as a grave concern. 'The appearance of the modern surface – smooth, white, shiny, sanitized – offers sufficient guarantee of protection from disease. An architectural soothsayer, or even a concerned homeowner, can point to a clean tub as evidence of a healthy future' (Braham 1997, 217). A safe family home[16] is no longer *just* about the eradication of germs and disease; it is also about subject demarcation by gender. We must see our own image in the receptacle – it should be that clean and mirrorical. 'The subject who looks at this sink [receptacle or urinal] is the phenomenological, self-conscious subject: the sink itself becomes a mirror in which "I see myself seeing myself" ' (Lahiji and Friedman 1997, 37). The mirage should cultivate an illusion of absolute subject integrity in the domain of gender, and the subject should be unencumbered by abject body fluid.

> Bathroom finishes must resist the accelerated tendency of matter to change state under the influence of water ... the glazed surface of the tub and of the tiled walls in the standardized room of fixtures is unchanging, or nearly so, requiring little of the regenerative maintenance demanded by other materials. (Braham 1997, 219)

The object constancy of bathroom fixtures is designed to quell anxieties about bodily ego boundaries and their instability. By appealing to an obsessive fantasy of extermination and removal – no part objects (floaters) in the toilet bowl – the lavatory caters to a modern individualist and puritanical wish to be unencumbered by the other, his/her shit and residue.[17]

It must be remembered that, because the 'rules of cleanliness were previously the province of religious doctrine' (Braham 1997, 217), and because such inscriptions relied upon whiteness as a trope and emblem of purity, there is, in the present-day manifestation of the hygienic superego, a compulsion to exterminate (by oversanitization) that which is not white (coded unclean or abject), a compulsion reminiscent of the older, Christian practice of ablution.

> Holy water doesn't run; its job is to resist loss, whether by drain, sin, or evaporation. Anointing and baptism originate in practices related to strengthening life, to the human seed (which chrism imitates), to fecundity and virility, also to extending the life of the soul past death. Bathing and oiling the body imply not just surface cleanliness, but replenishment through penetration to the body's depths. Life is liquid. Religious ablution is meant to pierce the skin on its way to the soul ... twentieth-century soul-saving requires twentieth-century plumbing. (Lahiji and Friedman 1997, 36–7)

The hygienic superego, with its appeal to medical science, epidemiology, and disease control, now supersedes religious teaching and pastoral ethics proper.[18] The hygienic superego is rationalized and inculcated by science in the secular age. But, like Christianity, the hygienic superego is beholden to the dialectics of purity and impurity – life and death – which are, at least in some moments, fantastical and illusory.

The ethical imperative of white, angelic purity (cleanliness) has always been an ideal which nobody can live up to. The gap or misrecognition between the body as it is and the body as it should be, between the real and the mirage, is orchestrated by restroom designs. The subject is intercepted at the vanishing point or plumb line. But the hygienic superego is not just about optics, colour, form, and texture. It is about subject stasis. 'Hygienic tales have likewise been used to explain the powers of the clean, white tub: the absence of colour and texture equals the absence of germs. The surface of the modern tub is frozen in glass, resisting change or inhabitation' (Braham 1997, 209–10). Just as the other is confused with the self in the mirrorical return, dirt and cleanliness are not distinct but rather co-dependent states.

> Clean and unclean do not exist in real opposition as two positive facts. Rather, they are two poles in a relationship of logical contradiction. The unclean is not a positive entity; it is only the lack, the absence, of clean ... the clean body always comes with a remainder, an 'excrement.' (Lahiji and Friedman 1997, 36)

Likewise, the masculine and the feminine – their subjective positions, identitarian coordinates, genital zones, and corporealities – are not oppositional so much as they are contiguous, or perhaps mirror-like, inversions or projections of that which has been driven underground or foreclosed in the other. Toilet choreographies and cleansing rituals demarcate the body so that sex can be anatomized and juxtaposed onto otherwise divergent and volatile flesh (Grosz 1994). There is nothing natural or given about the gender dimorphic positioning of bodies in the cultural landscape.

Bodies, like pipes and genitals, leak. Orifices obscure imaginary boundaries and psychically invested fantasies about impermeability. Fluids are unfaithful and promiscuous. We question and get upset about their whereabouts because they give us away and reveal others to have been where we ourselves wish to be or, conversely, where we do not want to be. Fluids escape the body and thus resist mapping.[19] Abject and unruly fluids upset gender. Fluids, like odours, threaten to overtake the primacy of sight in the modern optical arena, thereby obscuring body coordinates that are consolidated by the eye's exacting dissections. The 'flows' confuse body boundaries, and the disorientations are met with disgust (and sometimes desire).

Urine, menstrual blood, faeces, saliva, semen, and female ejaculate all threaten to alter the territory of the body, what Kaja Silverman (1996) calls the proprioceptive ego or sensational body (discussed in chapter 4). When the limits and contours of the body are uncertain, our relation to the signifying chain (Symbolic law) is unstable. Kristeva argues that excrement and menstrual blood are the two main polluting objects in phallocentric cultures. 'Excrement and its equivalents (decay, infection, disease, corpse, etc.) stand for the danger to identity that comes from without: the ego threatened by the non-ego, society threatened by its outside, life by death. Menstrual blood, on the contrary, stands for the danger issuing from within identity (social or sexual); it threatens the relationship between the sexes within a social aggregate and, through internalization, the identity of each sex in the face of sexual difference' (1982, 71). Both trajectories (from without and within) threaten gender identities in public toilets. Body fluids left in and around the white, oval, porcelain toilet bowl or urinal are met with disgust because they interfere with our internalized body maps.

Gendered anatomies are hard to decipher when they leak and smell out of place, time, and libidinally invested body coordinates. In her discussion of the 'mechanics of fluids,' Luce Irigaray (1985) notes that 'Fluid – like that other, inside/outside of philosophical discourse – is, by nature, unstable. Unless it is subordinated to geometrism, or (?) idealized' (112). The body's fluids symbolize disorder unless they are funnelled or plumbed (down the drain), subject to organizing spatial units (medical or scientific grids or geometrical maps that order and isolate fluids), or revered (as when menstrual blood was designated sacred because of its relationship to fertility in goddess worship). Irigaray also notes that fluids transgress and confuse boundaries that are integral to what science takes to be 'real.' In other words (her words), fluids defy the 'proper order' of the Symbolic and 'in large measure, *a psychical reality* that continues to resist adequate symbolization and/or that signifies the powerlessness of logic to incorporate in its writing all the characteristic features of nature. And it has often been found necessary to minimize certain of these features of nature, to envisage them, and it, only in light of an ideal status, so as to keep it/them from jamming the works of the theoretical machine' (ibid., 106–7).

We may understand what Irigaray refers to as the insistent inattention to fluids as an example of a phallocentric intolerance for the feminine coded as maternal and abject. As she says, 'Since historically the properties of fluids have been abandoned to the feminine, *how is the instinctual dualism articulated with the difference between the sexes?* How has it been possible even to "imagine" that this economy had the same explanatory value for both sexes? Except by falling back on the requirement that "the two" be interlocked in "the same" ' (116). If the feminine (or, for Irigaray, a 'sex that is not one') is absorbed into, or seen to be an inferior version of, the masculine; then it stands to reason that fluids marked as 'not coming from man' (the one and only sex) are most susceptible to abject horror. In other words, if menstrual blood in the phallocentric or cissexist economy is 'not male,' then it is the excess or remainder. Not only is menstrual blood abject, but it poses a noxious threat to non-trans heteromasculinities and to dominant ideas about an absolute and unchanging sexual difference between male and female. This is particularly the case when blood flows from bodies identifying as male or as masculine. Menstrual blood is, perhaps, along with faeces, the most culturally shameful bodily substance in those Canadian and American public cultures that have been influenced by Christianity and phallocentric reasoning. It is certainly subject to intense corporate-driven sanitary rituals.

A significant number of non-trans women and trans men interviewed are upset by the taboos surrounding menstruation and the shame those taboos provoke. Menstrual blood is subject to discipline by way of sanitation – for example, through corporate-driven advertising campaigns aligning menstruation with contaminate and pollution. Menstruation denotes a stain (blood red) upon life and codes the subject as impure (or as a harbinger of death). It should also be noted that undergarments are usually white because they are worn close to the body and meant to reveal abject body fluids. White panties, for example, show up menstrual fluids meant to be plugged up in a vaginal opening and absorbed by 'feminine hygiene' products (usually tampons). White light and clothing make menstrual blood ultra-visible, or, by contrast, demonstrate its absence (invisibility) or vanishing point.

Revulsion and denial work in concert when it comes to menstruation. This paradoxical response is evident in advertisements for 'feminine hygiene' products. As Emily, who is intersex and a transitioned woman, says, 'It's just straight back to body shame ... Look at the commercials for menstrual products ... Have you ever known anyone who had a blue period? ... It's absurd. It's this incredible over-sanitization.' Rohan, who is trans-masculine and butch, also focuses upon the ridiculousness of the blue-dye imagery used in advertising campaigns, the way in which it highlights cultural taboos surrounding menstruation and cultural investments in whiteness:

> Anything to do with menstruation [is taboo], like using blue dye in tampon commercials, I mean you can't use red dye? I mean we all know what colour blood is ... [Referring to tampons] And ... why are these objects white ... Like we know [about] ... toxic shock syndrome ... associated with after-effects left over in sanitary products from [the] bleaching processes to turn things unnaturally white, because then it's associated with cleanliness.

Zahara Ahmad also notices how the pedagogy of denial is employed in 'sanitary' product advertisements: '[menstrual] ... commercials are ... [intended] to [teach you] "How to hide that you have your period" ... [it is all about the] avoidance of its existence.'

Menstrual blood evokes shame (a revelation of the body and its insides) in public. Interviewees who bleed often say they are embarrassed by menstrual blood and by sanitary products, how to change or dispose of them. The disgust compels a will to absorb. In her definition of the words 'absorb' and 'absorption,' Molesworth (1997) writes, '*Absorb*: to take in without echo, recoil, or reflection: to absorb shock. *Absorption*: assimilation: incorporation' (76). The early French word for 'absorb' (*assorbir*) (or *absorbere* in Latin) meant to 'swallow up' or to 'suck in.' Menses are not funnelled back inside the body (although this may be a fanatical wish in our body-fluid-phobic culture), but into a negative (absorptive or non-space) without a visual trace or echo. The plumb line is blocked, and menstrual fluid shall be seen and heard to go nowhere. Molesworth notes, as well, that the capitalist logic and management of part objects (such as body fluids) is consumption oriented. The feminine incorporates by 'taking in' or 'absorbing' menstrual flows that are otherwise contrary to (or at odds with) a heteromasculine and typically phallocentric capitalist enterprise invested in the continual production of menstrual taboos.

Having to purchase and to change tampons, pads, sponges, 'divacups' (made of silicone), 'keepers' (made of rubber), and so on, in public space is described as disturbing or shocking (an antonym to the word 'absorb') to onlookers in the washroom, who are also embarrassed by the sight of menses. As Rohan notes, 'Women need to be changing sanitary, menstrual devices or whatever and of course [the assumption is that we should not] ... do [it] in front of other women ... it is coded as being very dirty and very shameful and ... something you ought to keep private.' Carol, who is a non-trans woman, emphasizes the cultural focus on concealment and how people are shocked by menstrual blood in the sink:

> I started using menstrual sponges instead of Tampax, for a while. And of course, when you're using a menstrual sponge, the big challenge becomes what you do when it's time to squeeze it out ... And so I got to the space where I would squeeze out my menstrual sponge in public restrooms ... in the sink [area] ... And ... I remember at least one time, having an old lady get very big-eyed next to me and seeming a little terrified.

Chloe, who is a non-trans queer femme, makes the same point about onlookers:

> [We are supposed to] dispose of ... sanitary products in a decent, unobtrusive way that almost pretends we don't menstruate ... [With respect to the 'keeper'] you *do not rinse your keeper in full view of other women!* ... Do not let other people know that your menstrual blood has just gone down the sink!

A third interviewee summarizes: 'I have a keeper and ... the whole process ... [of cleaning it creates] anxiety in, like, bathrooms, especially ... if I want to, like, dump it out in the sink, forget it! Like, that's a lot of stress, [it personifies] ... a leaky [feminine] body writ large.'

Trans-masculine interviewees who bleed are uniquely anxious about how to dispose of menstrual products. Menstrual blood is, as discussed above, coded as feminine and aligned with abject interiority. The architectural coding is especially obvious in the lack of menstrual disposal facilities in the 'men's' room. Lana, who identifies as a non-trans, genderqueer, femme gimp, and who often uses the toilet with transmasculine personal assistants, says, 'I've noticed that when I've been bleeding in the men's room ... [there is no]where to put your pad, so it's like "Okay, what do I do with this?" There's no trash cans in their stalls. So, I had a moment where I was like "God, I'm sharing a moment with trans guys." ' The problem is confirmed by KJ, who says,

> I worked with a trans guy with a full beard and everything and he's not on hormones anymore – he was on T [testosterone] – and so he has his period. But then when you go into a male washroom and you need to get rid of whatever, sanitary napkin, wrapper, tampon, whatever – you know ... How do you do that when the receptacle is obviously out[side the stall]?

Callum, who is a trans man, confirms that it would be 'traumatizing to have to come out of the stall and access it [the tampon machine] and go back in. So I was always very prepared [and brought my own product into the 'women's' bathroom].'

As indicated by the interviewees, the gendering of lavatories is painfully obvious in the presence or absence of what are called 'feminine hygiene disposal facilities.' The trouble with blood and the visibility of menses, in the 'men's' public toilet in particular, is not only that they confuse cissexist body maps attributing menstruation to female and feminine bodies exclusively but that they are at odds with a late-modern cultural imperative to absorb, or perhaps to contain, the liquidity of blood. As Foucault notes in his discussion of the symbolic of blood in the eighteenth century, blood was worrisome because it was 'easily spilled, subject to drying up, too readily mixed, capable of being quickly corrupted' (1978, 147). In other words, the bio-political regulation of the gendered and sexual body once secured by recourse to laws driven by anxieties about the unauthorized crossing of blood lines, those laws governing marriage, reproduction, kinship, and citizenship in particular, is now secured by attention to gender and sexual purity. Feminine 'hygiene' is not just about health and safety but about purifying and consolidating sexual difference, 'controlling our unruly pluralities [in the case of trans people],' as Hale (2009) notes. Gender incoherence is supposed to be eradicated by the internal grammar of the toilet.[20] Economies of power once focused upon the symbolic of blood are now, in the Foucauldian story of sexuality, more often consolidated by the deployment of sexual purity through discourse. The precariousness of blood – its troubling menstrual flows and dark red stains – reveals insides out. The spectacle is interlinked with a worry about death as represented by the archetype of the devouring maternal-feminine (discussed above), and also signified by the king and his capacity to have one beheaded. But this threat is no longer about human mortality, death, and dying, as governed by the older regimes of power heralded by the king, or even about a threat of disease and contagion. The worry is now about gender incoherence. Gendered ways of being at odds with a coordinated system of normative signs and significations governing the border between male and female, masculine and feminine, are related to larger worries about white heterosexual reproductive futurity (Edelman 2004) and the health of the nation.

Gendering Stink and the Mechanics of Disgust

> Odours and the act of smelling suggest a more personal and intimate identification with the other. In contrast, visual and auditory experiences are seen as more alienating acts than those of smelling, tasting, and touching. In the former experiences, the self does not consume or take in the stimulating particle. (Largey and Watson 1972, 1031)

Smell is implicated in angst about urinary and menstrual spills. 'Smells do not respect walls or national borders: they drift and infuse and inhabit' (Marks 2000, 246). In modern Western architectural designs, priority is placed upon the visual sensory apparatus. What Laura Marks (2000) calls 'certain [generic] sensory organizations ... entail not only an increasingly visual, specifically optical and symbolic, world, but also the abstraction and symbolization of all sense modalities' (244). In 'generic places' – such as public toilets in airports, hospitals, malls, and so on – odours are subject to intensive sanitization and purification regimes. Organic substances and chemical cleaners intermingle, but the latter are employed to extinguish the former. Body smells do, however, linger in the toilet and in memory. There is, as Marks suggests, a learned response to smell (205). She also writes that olfaction 'awakens deep seated, precognitive memories. Memories of smell endure much longer, even after a single exposure to an odor, than visual or auditory memories' (ibid.). Scents are not only uncanny (carrying with them associations both familiar and foreign), but sensual, erotic, and relational.

People perform aversions to smells emitted by the so-called opposite sex to secure identity-based differences. Gender is partially secured by enacting disgust in time and pace with modern architectural tempos. Refusing to lay claim to, or dissociating oneself from, a given scent is not only about the deadening of a given sensory experience in late modernity, or even the learning of disgust. It is about a will to consolidate obsessive claims to individuality, absolute corporeal integrity, and gender purity. These individual performances involve not only cleansing rituals, sanitization, and soap and water but also, increasingly, preferences for generic and chemical substances: the 'non-smells' of the modern toilet. Non-smells are generic, increasingly global, chemically based, and artificial emissions. 'Smell becomes unnameable. Beautiful smell becomes an absence that arises from the elimination of odor, part and parcel of the individualization and privatization of waste' (Laporte 2000, 65).

Gender purity is about the removal of promiscuous or otherwise wafting (unmappable) faecal odours that undermine subject-object distinctions. We are never entirely sure where a given fart or, to be more precise, emission of intestinal gas, may have originated (from whose anal region, mine or yours). Beautiful bodies are sanitized and well-plumbed bodies (with tight sphincter control). The oversanitization of public spaces, and of toilets in particular – coupled with an acute architectural focus on optics, whiteness, and acoustics – is to a large extent about securing gender integrity for the heteromasculine and cissexual subject, its purity and legibility in the public domain. We perform olfactory revulsion in accord with gender. But we do so in white, heteronormative, and cissexist spatial designs that are at odds with the smell and exhalations of people who are flesh and mortal. As Diane, who is a non-trans woman and a lesbian, elaborates, 'I think that we have an unhealthy desire that people don't smell like human beings and that they are, you know, that they don't touch one another.' Gypsey, who is a transgenderist woman, comments that 'Our sense of smell affects how we perceive a person.'

The ways in which olfactory sensations and fluids are managed in the lavatory is often heteronormative. As Jacq, who is genderqueer and butch, says, 'I've noticed condom machines [in bathrooms and] ... beside is the perfume machine. Which, you know, is kind of saying, "Hey, smell good and go fuck!" ' Jersey Star, who is white, of European descent, living in Toronto, and a non-trans queer woman, makes a similar observation: 'You walk in [to the bathroom] and there is the condom dispenser right next to the perfume dispenser right beside the maxi pads. That's all you need right there ... you got to smell good cover it up and get laid.' Body fluids and scents are, as suggested by interviewees, absorbed and sterilized in ways that are authorized by heteronormative and cissexist body politics. When covered up, or properly sanitized, body odours may be legitimately aphrodisiac or fragrant. When they are undomesticated and promiscuous or queer, they are considered foul.

The gendering of stink is one of the primary ways through which phallocentric and cissexual body maps are built into toilet designs. But it is not only the gendering of bodily exhalations that secures these body maps. How one strives to conceal, to perfume, to eradicate odoriferous body emissions is an indicator of how gender is disciplined by toiletry habitus. As Chloe says, 'It's an unwritten rule of femininity, right? You don't smell up somebody else's bathroom, or smell up a public bathroom.' A queer male interviewee similarly notes that 'women have been taught ... [to be] ashamed of bodily smells in a way that men don't [feel ashamed].'[21]

As argued above, the hygienic superego is most exacting in its focus on the feminine as opposed to the masculine subject. Consequently, feminine gender identifications are choreographed by a disproportionate avoidance of excremental smells and a heightened response to what counts as male stench. Chloe expresses disgust at masculine smells:

> I haven't been in a lot of men's washrooms ... [but] I'd say they smell. They smell bad, they just seem to me to be gross ... what makes them smell is men and ... their general untidiness ... My god, that smell! I mean, everybody knows that smell ... that urinal disgusting smell. Yuck!! ... I also don't know many women who want to go in the guy's washroom ... it's from all accounts, stinky and smelly and horrible.

The smell response marked by disgust, desire, or indifference orients, or is seen to be, a pivotal marker of gender. As Rico, a non-trans gay man, contends, more social significance may be given to the gendering of smells (and to modes of intolerance) than is actually warranted by the odours themselves: 'I actually heard a lot of people say ... "Oh, the men's bathroom is gross, don't go in there." But actually, I find the men's bathroom to be better than the women's bathroom a lot of times now. It's just, to be honest ... a different smell.'

While a feminine hygienic superego may be alarmed by wafting urinary smells in the lavatory, masculine people who use the 'men's' room often consolidate their gender identities by performing imperviousness to urinary stench. But not all masculine people (trans or cissexual) mimic olfactory indifference. As one trans male interviewee comments, 'I don't think men complain about grunge and stink. I don't know why that is, perhaps there's a sense that, you know, that's manly or something, but I don't think so. I think it's unhygienic and disgusting.' Ivan, who is queer, complains, 'I hate ... [the] smells, I have to say. But, visual stuff doesn't gross me out as much as smell.' Gypsey, who lives as a woman, identifying as transgenderist, concurs: 'I'm very sensitive to smell ... that smell ... in the men's washroom is absolutely offensive. It bothers me now [that I use the "women's" restroom], it does.' Jacq explains how as a genderqueer person you have to make direct contact with the interface of the toilet seat in the 'men's' room and how this proximity to urine, and to its smell, is horrifying: 'You're not going to be able to go in and pee on the ... urinal ... so you're going to have to sit on that horrible toilet, and I just think of ... [men who have] ... pissed or even worse [on the seat], you know ... So ... they do smell.'

While allegations of smelliness in the 'men's' room may seem trite, self-evident, or obvious, they are, in fact, shaped by long (and often forgotten) histories of olfactory intolerance in British and European cities (and elsewhere). Dominique Laporte (2000) contends that the history of senses – the degradation of smell in the modern era in particular, corresponds to the 'passage from promiscuity to modesty [which] cannot occur without a refinement of the sense of smell that entails a lowering of the threshold of tolerance for certain odours' (38). Smell gets in the way of vision (Laporte 2000). The degradation of olfactory sensations in the modern era and a corresponding disgust with the smell of the body's urine is indicative of an unprecedented sanitization of the body's aromas and secretions in public.[22] It also indicates a heightened interest in securing body boundaries from the 'common,' 'queer,' and/or 'perverse' odours of the 'other.'

Lingering body scents gender the toilets of today in ways that are upsetting to the senses. While masculinity is often marked by imperviousness to smell, the feminine is marked by a relative susceptibility to malodorous residues. Feminine odour is to be covered up (by perfume), cleansed (by soap and sanitizer), or dammed up (plugged at the point of emission). According to Tulip, who is a non-trans queer femme, 'There is something taboo about doing very natural things ... there is ... some sort of fantasy ... [about how] women don't fart, women don't burp, women don't do these things ... [gender is] so washed over.' Zoe, who is a non-trans woman, says that women perform disgust so as to distance themselves from others who are more likely to be aligned with that which is disgusting or non-normative.

> [We are] supposed to exhibit [disgust] in the bathroom [it] is almost us saying, 'Look, I know I'm disgusting, but that's even more disgusting.' So ... it's taking that focus off myself, as I know I can never live up to this ideal [feminine] norm that never shits, never smells, never blows my nose, or whatever, but if you can somehow defer that on to somebody else that's more disgusting then it takes [attention] away from my own inability to meet ... this mythical norm.

Disgust, as Zoe suggests, is not only identity bound but defensive and projective. Cissexual women sometimes perform disgust with aggressive and transphobic exuberance. One trans male interviewee says that women who are not trans yell at him in the bathroom: 'You're disgusting. You don't belong here.' Manifestations of disgust tell us something about how individuals consolidate gender and maintain subject integrity. 'Disgust responds to an encounter with something experienced as outside the self. That "Other" is felt to be noxious and ready to transfer noxiousness to the self. Therefore, one wants distance from the bad "Other." Disgust thus involves jeopardy to the self, which responds to that danger by devaluing – even despising – something outside, and determining to keep free of it' (Susan B. Miller 2004, 13).

Disgust is, as mentioned above, a transphobic defence. Rohan explains:

> Part of [gender-based regulation] is tied up in disgust ... I certainly see this in people's faces ... people find me disgusting in bathrooms ... They look at me ... with [a] very visceral sort of like 'ew.' Right, 'cause 'I can't read you, I don't know what you are' ... I've heard other trans people and queer people say that ... sometimes people respond to them with fear ... verbal abuse ... But often it's disgust, and you see it on somebody's face ... I think [it is about] illegible bodies.

As Rohan notes, illegible bodies are sometimes regarded with disgust by cissexuals. Trans and gender studies have focused upon how 'illegible' or 'incoherent' bodies in the domain of gender are often subject to transphobic injury (Butler 2004, 2005; Stone 2006; Noble 2004, 2006; Hale 2009). As interviewees indicate, these transphobic injuries are sometimes laced with a mode of address interspersed with disgust. Cissexist rules of gender purity set parameters upon who can be recognized as man or woman, as masculine or feminine, and so on, but interviewee lives exceed these parameters as they forge new ways of being gendered in the normative landscape. Gypsey says, for example, that she, as a transgenderist, 'hasn't had surgery but maybe to biological women that's a dirty mess.' Meredith, who is from Shanghai and Hong Kong and living in Toronto, a non-trans woman and queer, notes that those in transition are sometimes regarded by non-trans folk as 'less clean and less pure, less easily rationalized and understood.' As Shani Heckman, a white, non-trans genderqueer, butch living in San Francisco, says, 'I think that people just have a general idea about what the perfect human and perfect gender looks like, and if you don't' fit that, no matter what ... for some people that might be considered gross.' s. applebutters, a non-trans queer man, says, 'I don't personally ever feel unclean or anything like that. But, I think we are supposed to be made to, I think they, I think a lot of the tactics of the right wing and stuff like that are meant to make us feel totally dirty about ourselves.' Madison, who is non-trans and genderqueer, hypothesizes that 'there's something about not sticking to your gender that makes you seem ... unclean, or improper.'

Touching the Plumb Line

> A hand that touches is, in contact with the other, simultaneously an object touched. (Vasseleu 1998, 26)

> There is no denying the contact involved, but no means of objectifying the experience. Touch is the sense most affected by its object. (Vasseleu 1998, 100)

While vision and acoustic registers ensure the exteriority of the object, or, rather, its distance from the perceiving subject, tactility (like olfaction) implicates the self in the other. Touch resists subjective closure. It is an interface. Tactility folds 'opposites together so that they are also mutually reversed. This double enfoldment or invagination is the mutual "re-pli-cation" [*plier*: to fold] of differences in each other ... The identity of differences is never established in this doubling/interplay of meaning' (Vasseleu 1998, 33). Touch ignites an intimate reversibility that does not enable objectification or separation. People say they lose themselves in the touch of others. The caress implicates one in the other. While this interface is sometimes pleasurable, it can also be a source of displeasure when it interferes with subject integrity (ego boundaries and body borders). People are particularly troubled by contact with base body matter. Urine and faeces are expelled from the inside and therefore unnerve us as we feel our insides outside the body. Touch is also gendered. The feminine is, for example, culturally aligned with abject zones, surfaces, and fluids. Toilet designs imposing on women to sit put them in direct contact with abject remains. As Rico notes,

> Biological males don't necessarily have to make direct contact with their environment ... [Men] don't have to totally undress, or they only have to unzip, and ... there isn't as much physical contact with the environment ... But [women] ... actually have to make physical contact with ... a toilet seat or with other objects.

He further elaborates upon the relatively strict prohibitions on touch in the lavatory:

> You don't want to encounter other people's fluid ... [we] fear ... [other people's] fluid ... I mean most people ... are afraid, or think that [fluids] are gross ... I mean culturally ... body fluids can mean a lot of different things ... I think we're just such a society that's scared of the body and anything that comes out of it, it's scary to everyone.

Madison explains: 'It's also a space where people don't want to touch each other ... you never want to touch anybody in the bathroom ... [bodies] become filthy [in bathrooms].' As one non-trans gay male interviewee reflects, 'I think ... some people really, really, really dislike touching other people, viscerally ... bathrooms become a kind of battleground."

Sanitation and personal hygiene rituals guard against the effects of touch and close proximity to others. The rituals depend upon the metaphorical use of shit, disease, and contamination, as linguistic tropes, to divide the self from the other (coded as abject).[23] As Rohan intimates, our own shit never stinks: 'You don't use a disinfectant wipe to scrub your ass off before you sit down, you scrub the toilet seat. The assumption is that my body is clean. That person's body ... not so much.' The focus upon personal hygiene and sanitation is, thus, projective and defensive. Excremental filth, body soils, and foul smells are felt to be 'foreign' and external to the self. LGBTI people vulnerable to expressions of disgust are hyper aware of their own susceptibility to faecal rhetoric. Some interviewees submitted to highly intricate cleansing and purification rituals in order to avoid being associated with abject body fluids. As Jay, who is genderqueer and gay, explains,

> When I go into the bathroom ... I wipe off the seat, I am very aware of, like, who saw me come in here who doubted that I was supposed to be in here and who is watching me not sitting down right away ... that ... [is] a conscious thought every time I go to the bathroom and [I] have to wipe off the toilet seat.

Chloe says, 'I'm the person who carries my own lavender soap. I have my own sanitizer, and will grab a paper towel and hold it in my hand to open the washroom door.' Velvet Steel, who is a trans woman, also purchases cleaning products for personal use in public lavatories:

> I always buy my travel wipes at Shoppers' Drug Mart. I always keep those with me ... I always give a wipe to the toilet before I actually sit down because I *will sit*. I will make sure that I wipe that toilet seat good and clean, use the towel to wipe it down dry, then [I] sit down. Or, I will ... put toilet paper across it ... but I also give myself a wipe – with a *fresh* one.

Nandita, who is a person of colour from Pakistan, now living in New York, and a non-trans bisexual female, says that she cannot help feeling dirty in the bathroom:

> I'm very cautious about cleanliness ... I tend to have ... two, or one if I'm in a hurry, strips of toilet paper around ... the frame ... or the seat. And then I sit and then I still feel dirty. I just feel dirty in a public bathroom. I just feel dirty whatever I'm doing. Even if it's changing clothes, I feel dirty ... I just feel dirty about being in the space.

Rico, who is a gay man, is loath to touch the toilet-seat:

> When I walk into a bathroom that's not in my own home, I have this feeling that I don't want to touch anything. It kills me to have to sit on a toilet seat that strangers use ... they pissed all over the seat, wiped faeces on the toilet ... They miss somehow ... and I have to go where they go and that disgusts me ... I'll use the paper towel to open the door ... whatever they've touched is on what I have to touch to get out [and that upsets me].

To butt up against, hit, converge upon, accidentally caress, or feel the inner perimeter of the bathroom (or another in it), the metallic, porcelain, and tiled apparatus enclosed within, is to be (or feel) contaminated, as interviewees explain.

Fears about contamination are mirrored by architectural designs that espouse an impossible ideal of cleanliness. Images of a racially specific, white aesthetic are market driven and orchestrated by institutionalized toilet designs. Ideas about urine, germs, disease, and contamination upon contact gender people as familiar or foreign, as sovereign or disease ridden, as safe or infectious. David, who is a non-trans gay man, even suggests that toilets are invested with subjectivity: 'Some public toilets ... are not so clean or ... of questionable character.' Layal, who is Arab and living in Toronto, a non-trans queer femme, pinpoints the troubling equation: 'Public equals dirty.' Sarah says that toilets are 'white-washed' because people 'want to fake the illusion of clean.' With great insight, Callum states, 'I always wonder what that clean word means anyway.'

Inhaling Gender and Haptic Visuality

> When we find there is nothing to see, there may be a lot to feel, or to smell. (Marks 2000, 231)

People are gendered, classed, and racialized as unclean by disciplining and training the senses. Visual intolerance for 'dirt' is dependent upon what Laura Marks (2000), in her study of intercultural cinema, calls 'haptic visuality,' which she defines as a modality of vision that 'functions like the sense of touch' (22) or that incites smell and taste. Western modernity privileges vision (and, to a lesser extent, the acoustic) and downgrades olfactory and tactile sensations. Vision is often employed in the West to create and maintain distance between people because it is the 'sense generally most separate from the body in its ability to perceive over distances' (ibid., 132). Olfactory sensations, by contrast, are evidence that 'molecules ... have reached the membranes within our noses. Smell requires contact, molecules coming into touch with receptors' (ibid., 113–14).

The modern lavatory is designed to separate people by gender and to remove the body from its urinary and faecal remains. Olfactory and tactile remainders (leftover scents and fluids) are objects of hygienic concern. They are, thus, sanitized and removed. The public toilet is dually structured by an optical design enabling us to visually separate (objectify and fragment) bodies, *and* a haptic design inciting uncanny recalls and affiliations. In other words, the hyper-attention (however phobic) paid to body fluids and smells ignites disgust (predicated upon separation) *and* desire (a relation or mode of affiliation resistant to categorization). It is not coincidental that most expressions of disgust are in response to fluids and smells, because, as Marks contends, 'Tactility [like smell] cannot be a distance sense' (2000, 132). Touch demands proximity, which is typically offset by modern architectural designs that accentuate sight and sound. What we might call a haptic architectural design troubles identifications that are predicated upon absolute incorporation or excorporation. Haptics invite one to affiliate, to touch, taste, smell, and/or caress without an absolute investment in sameness (incorporation) and difference (excorporation). A focus upon olfactory sensations and tactility in modern architectural design can undo and upset cissexist gender codes. This is because our relation to people culturally coded as other through heteronormative and transphobic logics, racism, classism, and other imperial and modern systems of exclusion is unstable. We never know precisely who or where we are in relation to others, and so normalizing identifications are, to varying extents, architecturally regulated.

Haptic visuality, like touch, is a modality of seeing that is predicated not upon an alienating distance but upon a tendency to 'move over the surface of its object rather than to plunge into illusionistic depth, not to distinguish from so much as to discern texture. It is more inclined to move than to focus, more inclined to graze than to gaze' (Marks 2000, 162). Haptic visuality is less about identification and objectification than it is about a 'bodily relationship between the viewer and the image ... a dynamic subjectivity between looker and image' (ibid., 164) characterized as a co-presence and a feeling of being 'nearby.' According to Luce Irigaray, touch is not only the first sensation experienced by the foetus but a modality through which one is connected to an exteriority (composed of people and things) which language and modern visual technologies (the mirror, the camera, and so forth) later turn into objects. Marks suggests that haptic images are

> erotic in that they construct an intersubjective relationship between beholder and image. The viewer is called upon to fill in the gaps in the image, to engage with the traces the image leaves. By interacting up close with an image, close enough that figure and ground commingle, the viewer

relinquishes her own sense of separateness from the image – not to know it, but to give herself up to her desire for it. (Ibid., 183)

As with haptic images, tactile and olfactory senses call for erotic engagements. Promiscuous fluids and smells disorient gender. The disorientation is an effect of proximity, as opposed to a voyeuristic separation between subject and object. Haptic disorientation enables neither identification (incorporation) nor disidentification (excorporation). Marks suggests that haptics might 'corroborate a kind of visuality that is not organized around identification, at least not identification with a single figure, but that is labile, able to move between identification and immersion' (188). An erotic haptic takes pleasure in an affinity that bears no absolute relation (or unrelation, as the case may be). Close quarters requiring touch, or proximity to others by way of smell, can be understood in terms of the caress. As Emmanuel Levinas (1989) writes,

> the caress does not know what it seeks. This 'not knowing,' this fundamental disorder, is the essential. It is like a game with something slipping away, a game absolutely without project or plan, not with what can become ours or us, but with something other, always other, always inaccessible, and always still to come. (Quoted in Marks 2000, 184)

As in a sexual encounter, one can lose the self, but in a pleasurable way. To be undone by another is to lose perspective, or grounding, in an optic visual field that otherwise ensures identity-based differences. Toilets, differently configured and occupied, might work this way (as I suggest in the following chapter).

Shit and Anal Imaginations

> This culture has lots of taboo stuff about shit. (A trans male interviewee)

> As everybody can tell by walking through a newly completed building, contemporary architecture has a harmful smell. (Frascari 1997, 163)

Aversions to dung (abject par excellence) are gendered and gendering. Female gender identifications are often shaped by aversions to scatological remains, while male gender identifications are more likely to enact an imperviousness to scat (or use it to expand the symbolic territory of the body). The faecal-mass ejected may *feel* like a little penis (as hypothesized by Freud),[24] or an 'unfeminine' penetration or cut into the otherwise rigidly gendered geography of the toilet. It is certainly the case that interviewees who use the 'women's' room tend to be disgusted by shit,[25] and many abstain altogether from having bowel movements in public rooms. As Haley recalls, 'I remember being in a public bathroom in our school, and having a couple of my friends talk about how they would *never* poo in the bathroom. And they thought it was so disgusting ... when people did [eject faeces].' Frieda, who is of British heritage and living in Toronto, non-trans, and queer, agrees: 'There's this unspoken code of conduct in [the] women's [room] that you don't do number two.' To leave behind, to be seen to have made, or to be surrounded by the aroma of, stool is to trouble feminine gender identifications. Stool is to be invisible (flushed away) or, as Jacob says, 'for home.'

Some interviewees talk about angst-ridden moments when they needed to defecate, but could not, or did not want to. Others said that worries about transphobia and homophobia inhibit one's capacity to discharge faeces. In Chloe's words, 'I think hassles around homophobia ... Or ... for some [other] reason [LGBTI folks] can't be relaxed enough in a public washroom to move their bowels.'

If gender protocols in the 'women's' room demand that one avoid defecation as unacceptably messy, the rules governing defecation in the 'men's' room are almost entirely reversed. For those invested in dominant, cissexually defined masculine subject positions, there appears to be pride in, or space taken up by, faecal droppings. This pride is in direct opposition to the shame sometimes felt by interviewees who use the 'women's' room. Rohan elaborates upon the gender difference as follows:

> It's funny ... guys will take a dump and leave it [in the toilet bowl] and not flush; it's, like, 'Look what I left behind! Look what I did!' Like a little kid, 'Look what I made,' whereas women would never do that.

Callum also observes the gendered responses to dung:

> I think ... [there is pride around not flushing for men] ... like a 'Look what I made,' so everyone can see ... But ... then it's the same reason that you would be flushing [in the 'women's' room] because you are ashamed about it. And a woman would never not flush. Only if it's plugged, and even then it's traumatizing ... [and] horrible, like, you'll close the door as you leave.

Hegemonic masculinities seem to be authenticated by a willingness to perform imperviousness to cultural infection by faeces. One's own excrement is also to be made visible. In his discussion of sport and the territorial anus, Brian Pronger (1999) contends that white heteronormative masculinities are established through territorialization. He writes that there is, in dominant displays of masculinity, a 'colonizing will to conquer the space of an "other" while simultaneously protectively enclosing the space of the self, in an attempt to establish ever greater sovereignty of self and consequent otherness of the other' (376). The visual presence of the phallus is exaggerated while the anus is closed to curtail the threat of penetration and subsequent feminization. Faecal remains do not represent an open or gaping anus, however. They represent the markings of a phallic heteromasculinity that is committed to the usurpation of the other's body space. To shit in public space is to claim or mark the place as one's own. There is a none-too-subtle correlation between heteronormative masculinities and the territorialization of public facilities through the spread of dung. Evidence is left behind, so to speak. As Rohan comments,

> I used to have to clean men's washrooms and women's washrooms ... [and] the stuff that men do, like shitting next to the toilet on the floor. Smearing it on the wall ... I've never cleaned a women's change room where women were like sitting there playing with their poop, or doing performance art with it or something.[26]

Of course, not all masculine identities are secured (or territorialized) by the spread of shit. Gay and/or trans men in this study tend to take issue with public displays of excrement and ejaculate.[27] Interviewees' complaints about the stench in the 'men's' room may be a reaction to the aggressiveness symbolized by indiscreet and territorial droppings. Dung spilled onto floors and toilet seats or smeared on stall walls and doors projects the masculine (phallic) body into public space.[28] (This is in direct contrast to the way menstrual blood symbolizes feminine interiority and is absorbed or plugged up to relinquish a territorial and expansive claim to public space.) Space is acquired symbolically by the way those who are heteronormative and conventionally masculine sometimes spread the dark mass. The presence of shit 'out of place' may also signal trans and homophobic hate. If the anus and the homosexual are equated in the homophobic imagination, the performed repudiation or display of shit – wiping it on walls, for example – is very likely an aggressive repudiation of anal eroticism. 'That abjectifying – and therefore effeminizing – anality is a condition that homophobic masculinity repudiates by constructing it as the distinguishing hallmark of a recognizable category of homosexual person' (Edelman 1994, 169). Insofar as the anus signifies a gap or negative space – an invagination, so to speak – it cannot solidify non-trans heteromasculinities invested in phallic, penetrative capacities and a simultaneous refusal of vaginal tissue, when, as Freud tells us, the act of defecation *feels* like penetration. Edelman suggests that the 'anatomical "cavity" denoted by the "cloaca" ... [conjures up] anxiety of an internal space of difference within the body, an overdetermined opening or invagination within the male, of which the activity of defecation may serve as an uncanny reminder' (1994, 162). People less beholden to homophobic prohibitions may be more willing to avow an 'internal space of difference – an opening or invagination.' They may also be more able to deal with the symbolics of death and disorder signified by body matter 'out of place.' Heteronormative and cissexist displays of masculinity may be disproportionately dependent upon grandiose fantasies of absolute subject demarcation (from the maternal-feminine) and immortality.

Dominant, white heteromasculinities tend to present themselves as if they are invulnerable to the anus and to the messes it makes (coded as feminine and aligned with a queer interiority). As Pronger notes,

> Masculine desire protects its own phallic production by closing openings, preeminently the anus and mouth ... in short, any vulnerability to the phallic expansion of others. Rendered impenetrable, masculine desire attempts to differentiate itself, to produce itself as distinct and unconnected ... The point of this conquering and enclosure of space is to make bodies differentiate themselves from the vortex of unbounded free-flowing desire ... and thereby establish territorial, sovereign, masculine selfhood. (1999, 381)

In his discussion of the anus and the toilet, Lee Edelman (1994) makes a similar observation. He suggests that 'urinary segregation' is about the establishment of sexual difference. The phallus must be in plain view (before the urinal) to differentiate the masculine body from a feminine body – both of which, in the stall, experience a 'loosening of sphincter control, evoking, therefore, an older eroticism, undifferentiated by gender, because anterior to the genital tyranny that raises the phallus to its privileged position' (161). To shit in place is, for heteromasculine subjects, risky. 'For the satisfaction that such [intestinal] relief affords abuts dangerously on homophobically abjectified desires, and because that satisfaction marks an opening onto difference that would challenge the phallic supremacy and coherence of the signifier on the men's room door, it must be isolated and kept in view at once lest its erotic potential come out' (ibid.). Heteromasculinity demands not only that the anus be refused as a site of pleasure but that the subject's excrement be imposed upon the space of the Other (in and around the stall).

Gay and/or trans men in my study seem to be less likely to use faeces as a means to claim public space and to seal the territory of the body. They are also more willing to acknowledge how shit can humble the body. Gay men sometimes regard faecal matter as part and parcel of what it means to be human and thus mortal. Faeces are also by-products of anal sex.[29] As Ivan reflects, 'I'm a little bit embarrassed about crapping in public, just a little bit ... Audible stuff ... mainly it's the noise ... it's about the body being seen as low, abject and dirty.' He also suggests that the relationship between disgust, abjection, and faeces is very much about sex and how one is able to come to terms with the liquidity of the human body:

> as a gay man who has anal sex, you come across shit, right? And heterosexuals have anal sex too, so ... I'm not the kind of person who gets disgusted by these very normal, natural processes around sex ... Sometimes [sex] gets dirty ... So, there's shit on the condom. There's blood on the condom. There's vaginal fluid ... we're messy, liquid creatures.

Ivan also notes that the presence of faecal matter can sometimes bother him and make him lose his erection:

> sometimes it [shit] comes out when you have sex ... the smell of shit bothers me, so if there's fucking going on and there happens to be a lot of shit, then the smell bothers me, the smell will make me lose my erection, or it will stop the sex ... [But] a lot of gay men that I've been with treat it [shit] as a very normal part of sex ... I don't know about [what] other people's experience is, but mine has been [that] people don't freak out, at all, which is good.

If the homosexual (or sodomitical scene) signifies a kind of social and reproductive death, as Edelman (2004) argues in his discussion of 'reproductive futurism' in white, homophobic American culture, we may surmise that homophobic preoccupations with faeces reflect a discomfort with ways of being together that are felt to be 'anti-family,' foreign, or abject. 'The Freudian pleasure or comfort stationed in that movement of the bowel overlaps too extensively with the Kristevian abjection that recoils from such evidence of the body's inescapable implication in its death; and the disquieting conjunction of these contexts informs, with predictably volatile and destructive results, the ways in which dominant American culture could interpret the ' "meaning" of male-male sexual activities' (Edelman 1994, 161). Kristeva (1982) tells us that the abject (faeces in particular) upsets fantasies about subject integrity and is, ultimately, associated with death or annihilation. Interviewees also notice how the toilet is haunted not only by the ghost of homosexuality but by the spectre of death and dying. Rohan focuses on human frailty:

> it's that [human] frailty ... that idea that yes I am going to get old, my body is going to fall apart, and it's going to get progressively leakier, smellier, less controllable. I'm going to have less control over my bladder and my bowels as I age because that's just part of aging ... I think that bothers people. To be confronted with that bothers people.

Velvet Steel compares the bathroom to the hospital: 'I think most people see it [the bathroom as] ... a bad place to go. It's almost like going into a hospital ... people [worry that] you're never going to come out!' Another trans interviewee comments on faeces and organic decay:

> Faeces doesn't smell so good. It's like the smell of rotten chicken ... I think that they're waste products for a reason and decay is ... we find it abhorrent because we don't want to think about decay. We are all [supposedly] perfectly healthy and we are all going to live forever there is no decay there is no rot, there's no waste product. We are all just *clean as a whistle*.

Worries about hygiene, rot, and decay are intimately tied to gender. Whiteness is about abstinence and absence (no longer here), as mentioned above, and gender purity is about obsessive attempts to police the borderlands between male and female, masculine and feminine; each is preoccupied by a negative or tertiary space, a gap or disconnect between life (coded as masculine virility) and death (symbolized by the sodomite and the maternal or devouring feminine).

Curiously, the difference between the masculine and the feminine is managed by a puritanical, white, body politics that inhibits touch. The condemnation of touch and unlawful (usually public) sex is enforced to preserve a strangely virginal and pristine whiteness in a space that is oversaturated by heteronormativity. The toilet is a site of death and disease (not life), and it is supposed to be a site of orificial (anal) disavowal (not homosexual activity). Bathroom technologies police touch and keep bodies apart.[30] The designs and cleaning technologies displace concrete worries about death and dying onto psychically invested concerns about gender purity and sexual abstinence. Olfactory and tactile emissions are disowned, or perhaps deadened, while visual and acoustic registers are employed to enact a vanishing of excrement at the stillwater mark or in the pipeline.

Karl Marx wrote that modern alienation under capitalism is, in part, about an estrangement from the body's sensory apparatuses, with the possible exception of vision (and I would add acoustics). The degradation of touch and smell, what Marks (2000) refers to as the 'close' senses, fundamentally alters the social landscape. Marks argues that modern optics are predicated upon 'symbolic representation,' whereas tactile epistemologies (and the relationships they engender) give rise to mimesis. She defines mimesis as a 'form of representation based on a particular, material contact at a particular moment' (138). Touch conjures up a memory encoded in the senses. Mimesis also 'presumes a continuum between the actuality of the world and the production of signs about that world' (ibid., 139). So, for example, the automated sink that can be turned on without touch, without the turning and release of a valve, is an alienating optical design as opposed to a mimesis. The automated technology secures distance and prohibits touch. Summarizing the foundational work of Horkheimer and Adorno (1972) and Benjamin (1978) of the Frankfurt School, Marks writes, 'Mimesis, they argued, is a form of yielding to one's environment, rather than dominating it, and thus offers a radical alternative to the controlling distance from the environment so well served by vision' (2000, 140).[31]

There is also an uncanny aspect to mimesis because it conjures up relations now past but encoded in touch. In mimesis, one not only remembers (or senses what has been forgotten) but also yields to (does not objectify) the organic. Horkheimer and Adorno, for example, describe tactile epistemology as a 'yielding form of knowledge ... [not unlike] the death instinct, the willingness to merge back with nature' (quoted in Marks 2000, 143). This epistemology is distinct from what Horkheimer and Adorno refer to as a 'mimesis unto death' (1972, 57) in which factory workers, or bathroom patrons for that matter, line up and choreograph and sequence their movements in keeping with modern capitalist ethics governing productivity, utility, and efficiency (not to mention sanitation). What might be called an insatiable capitalist mimesis shelters and segregates the self from nature, from the immediate environment, and blunts the senses in doing so.

Modern bathroom designs, increasingly reliant on automated technologies, cultivate a mimesis that dulls and rounds out the senses. With the deadening of the intimate (less objectifying) senses we see a refusal to validate trans-identifications and queer sexualities that do not lay claim to a clean and private cissexist and heteronormative body politics. There is, as suggested above, a generic and anaesthetized spatial politics of the bathroom that incorporates any remaining sensory experience into a white, Westernized, privatizing, and reproduction-oriented heterosexuality. 'It seems reasonable to worry that as culture becomes globalized, sensuous experience is becoming both universal and placeless. "Non-places," or generic places (malls, airports [and, I would add, toilets]) are proliferating around the world … and would seem to bring with them certain sensory organizations' (Marks 2000, 244). A mimesis responsive to touch or smell is now subordinated to modern, generic optics and acoustics. Toilets are increasingly sanitized, automated, and reliant upon globally marketed apparatus. 'When it is separated from its source and packaged, smell becomes a simulacrum, the scent of nonplace' (ibid., 245).

Although there is a modern disdain for touch and smell in the design of the toilet, space is made – however provisionally – for the organic, the sensual, and the feminine by those who use it in unauthorized ways. 'Against the tide of the commodification and genericization of sense experience, pools of local sensuous experience are continually created anew' (Marks 2000, 245). Sex in bathrooms, erotic uses of and fixations upon fluids and smells, and excitations about who may be in a neighbouring stall based upon curiosity and sensual intrigue, as opposed to phobic and exterminatory impulses, make room for new (while recalling older) modalities of mimetic engagement that are typically defamed in modern, Western, capitalist, cissexist, and heteronormative organizations of culture. Eroticizing partition walls (and puncturing them to make glory holes); finding pleasure in the contravention of prohibitions placed upon public mixing; making or fantasizing about an unauthorized touch while seated upon the throne or standing before the urinal – these are all ways to create new modalities of desire and identification. To queer, or to sensate, modern restrooms we must support multiple ways of being gendered and their associated structures of desire in the domain of the social. In the following chapter, I consider how gender is disciplined by prohibitions placed on sex in public.

6 Sexing Gender: The Homoerotics of the Water Closet

A lot of the questioning around who belongs in which restroom always has a certain homophobic quality to it.

(Interviewee who is transgender)

[The bathroom] space is colonized by heterosexual men, so ... yeah, I monitor myself.

(Ivan, who is a non-trans queer man)

It's true, queers do fuck in bathrooms – we simply can't resist.

(Temperance, who is a non-trans queer femme)

Foucault (1978) argues that the history of sexuality in the West is about an incitement to discourse. Sexuality has not been repressed so much as it has been deployed. Talk about sex is an invitation to wonder about the body, its pleasures, capacities, and departures from norms authorized by disciplinary institutions and their secular apostles. The conjugal family imposed itself as a normative point of reference. What Foucault calls 'peripheral sexualities' (39) were entered into medico-legal and psychiatric lexicons. Homosexuality was invented along with other 'perverse' identifications itemized and studied by sexologists. Sex even made its presence known in the architectural designs and operations of the disciplines dating back to the eighteenth century.

> But one only has to glance over the architectural layout, the rules of discipline, and their whole internal organization: the question of sex was a constant preoccupation. The builders considered it explicitly ... What one

> might call the internal discourse of the institution ... was largely based on the assumption that this sexuality existed, that it was precocious, active, and ever present ... [It became] a public problem. (Ibid., 27–8)

The disciplines functioned to extract a truth about sex. Much about the individual was to be revealed by subjecting sex to surveillance, inspection, and diagnosis; the study of sex was the point of entry into the soul.

Part of what was to be discovered about sex is related to gender. As Jay Prosser (1998b) argues in his discussion of nineteenth- and early-twentieth-century sexology, 'what sexologists sought to describe through sexual inversion was not homosexuality but differing degrees of *gender* inversion' (117). Challenging Foucault's claim that the homosexual was invented with the discovery of sexual inversion (which, for sexologists, came to be a master signifier of homosexuality), Prosser argues that 'sexual inversion *was* transgender, and while homosexuals certainly number among inverts, the category described a much larger gender-inverted condition of which homosexuality was only one aspect' (ibid.). Not only were sexuality and gender conceptually entangled, but homosexuality overshadowed and came to represent transgender.[1] For example, Foucault says in his discussion of homosexuality that it was thought to be a 'certain quality of sexual sensibility, a certain way of inverting the masculine and the feminine in oneself. Homosexuality appeared as one of the forms of sexuality when it was transposed from the practice of sodomy into a kind of interior androgyny, a hermaphrodism of the soul' (1979, 43). Sexual practices such as sodomy were thought to reveal interiority, and the interiors of the modern subject were gendered. In this discursive economy, gender inversion signifies homosexuality. There is little or no room to theorize heterosexuality among gender inverts or to consider how homosexuality may be one way (among other ways) to signify gender inversion (or what we now call transsubjectivity). While I am in agreement with Foucault's claim that the idea of 'sex' 'made it possible to group together, in an artificial unity, anatomical elements, biological functions, conducts, sensations, and pleasures ... [and to] make use of this fictitious unity' (1979, 154), I take issue with the idea that homosexuality, as opposed to transgender, was the primary or exclusive focus of sexology. Certainly, the way the figure of the gender invert was deployed to signify homosexuality in sexology, and later in queer theory, functions to negate the specificities of trans lives, their gender identifications and sexed embodiments. I do suggest, however, that an absolute separation between gender and sexuality in historical and sociological inquiries may, in some cases, be erroneous. The regulation and bio-politics of gender and sexuality often work in tandem. Gender is deployed to secure what Judith Butler (1990) calls the heterosexual matrix, but this isn't to say that trans people cannot be heterosexual or that all cissexual heterosexuals are heteronormative and invested in the reproduction of dominant body politics. At the same time, gender and sexuality are not one and the same. There are significant differences in and between gender identity and sexual orientation.

Matrices of heterosexuality are mapped onto bathroom designs, where sex is deployed to secure a binary gender order. If gender polarities are secured by compulsory heterosexuality, with its commitment to the family and to reproduction, how does queer sex for pleasure in the public domain upset gender? Or, conversely, how do institutions of heterosexuality deploy the bathroom 'sodomite' to secure a bio-political hold on gender for the sake of the population, its health, well-being, and reproduction? Heterosexual majorities view sex in bathrooms as 'dirty,' 'immoral,' 'perverse,' 'unhealthy,' and 'criminal.' The affront to polite civil society has little (if anything) to do with public health and safety. In his discussion of bio-power (power over life), Foucault (1978) contends that technologies of sex depend upon the discursive '*themes* of health, progeny, race, the future of the species, the vitality of the social body' (147, emphasis added). Queer sex in bathrooms is subject to surveillance because it flouts the prohibitions placed on homosexuality. It also upsets the gender coordinates of public space that are predicated upon the homosexual injunction. By making alternatives to the norm public and visible, it brings to the surface (however tentatively) the disavowed content of heterosexual self-identifications. By this I mean to suggest that people committed to heterosexuality (as a norm, nature, ideal, and inevitability) are confronted (troubled and sometimes confused) by modes of desire (and gender identifications) they have already (without knowing it) refused. While I do not pretend to claim that sex in the water closet can overturn the structure and composition of heteronormativity, it can incite ways of being together, loving, desiring, and identifying that are at odds with the normative regime. Queer erotic cultures inaugurate new and 'changed possibilities of identity, intelligibility, publics, culture, and sex that appear when the heterosexual couple is no longer the referent or privileged example of sexual culture' (Berlant and Warner 2002, 187). Sex in the lavatory is desire out of normative bounds: it is messy, unproductive, licentious, epicurean, and hedonistic. The disciplining of sex in toilets is not only about the regulation of gender; it is about what Lee Edelman (2004) calls white reproductive futurism: an identification with white patriarchal succession and the reproduction of cissexist body politics.[2]

Sex can upset 'gender as usual' – particularly when it is public and queer. This is why so much attention is given to how, with whom, where, how long, under what conditions, and in what space sex can be had. Sexuality is a linchpin stabilizing gender – even when we are unclear about what precisely is being secured. Sexuality impacts upon the gendered bodily ego. To illustrate, Kaja Silverman (1996) notes that erotic zones are 'features of the bodily ego ... we all know that the areas of the body in which someone experiences sexual pleasure have a lot to do with his or her identity' (14). Butler's matrix of heterosexuality is, similarly, predicated upon a concord (however unstable and inessential) between the erotic and the gendered body as it is materialized through discourse, fantasy, power, desire, and so forth. 'The construction of stable bodily contours relies upon fixed sites of corporeal permeability and impermeability. Those sexual practices in both homosexual and heterosexual contexts that open surfaces and orifices to erotic signification or close down others effectively reinscribe the boundaries of the body along new cultural lines' (Butler 1990, 132). Anal sex, oral sex, water sports, the use (or materialization of) lesbian phalluses,[3] masturbation, coprophilia (arousal by defecating), hygrophilia (pleasure in body fluids), mysophilia (olfactory predilections for body discharge), and numerous other practices centre desire upon unauthorized orifices, positions, and fluids in ways that can upset normative body politics.

In short, gender can be undone, as it is also remade and subject to negotiation, by modalities of desire and sex play that expand the bounds of missionary sex. This is not to say that gender is determined by the sex one has, private or public, but that all sexualities are anchored to, or moored by, desire. Fantasies about the stability, nature, and essentialism of the body and its gender are often dependent upon a fixing of sexual practices: upon their confinement and docility. Heterosexual bifurcations of gender typically depend upon the confinement of sex acts to private spaces, to the reproductive and marital family bed, to 'respectable' partners (only one at a time) of the 'opposite' sex. Only certain orifices can be penetrated (usually a vagina or mouth) by a particular organ (usually a penis), with only certain body fluids ejaculated (usually semen); and all of this is to fit into a preordained choreography with very few exceptions or room for negotiation and play.

In their discussion of public sex, Lauren Berlant and Michael Warner (2002) write that the 'queer world is a space of entrances, exits, unsystematized lines of acquaintance, projected horizons, typifying examples, alternate routes, blockages, incommensurate geographies' (198). Public sex gives rise to queer cultures. Erotic bonds are forged out of wedlock and in spaces heteronormative majorities deem impure, unsanitary, and dangerous. To have 'dirty sex' (sex out of place) is to upset a white, national, heterosexual impulse to align the disenfranchised (the queer and/or transgender subject, those denigrated by racism, the disabled, the unemployed and street active, and so on) with abject faecal matter. To make something erotic of the grotesque is to override the revulsions set in place to discipline, domesticate, and dim the numerous ways we are not only gendered but sexual. 'National heterosexuality is the mechanism by which a core national culture can be imagined as a sanitized space of sentimental feeling and immaculate behavior, a space of pure citizenship' (ibid., 189).

Public sex prohibitions are about gender. But they are also about a puritanical will to discipline public cultures predicated upon the ephemeral. Because it is seemingly immune to the various prohibitions placed on the body and its sexual capacities, queer sex in toilets is an affront to the normative gender identifications that are sanctioned by Western modernity, Christianity, capitalism, and the white nuclear family as a national and colonial icon. Sex in public not only reorganizes divisions between private and public; it invites us to question the nature of gender positions solidified through sexual prohibitions. Body boundaries are precarious when orifices open and fluids shed. The spectacle of sex in bathrooms brings to light the disavowals upon which heteronormative and typically cissexual gender identifications are forged. When bodies intermingle and touch, caress, penetrate, and trade in culturally abject fluids, normative gender identifications are sometimes upset. Nothing is or remains as it was. The North American toilet is a cultural icon of disgust because it is a repository for all that is repudiated in the service of 'coherent' gender ego ideals. It is no coincidence that in Freud's discussion of child psycho-sexual development he refers to the toilet as a place of forgetting.[4] 'But with each raised and lowered seat, every splash of urine, every tear of toilet paper littering the floor, the bathroom and its plumbing point to the impossibility of keeping intimacy (the personal) out of the public, and of keeping the sovereign individual free of contamination ... paradoxically, plumbing also connects you to every other denizen in the communal rush to separation' (Morgan 2002, 176).

The Heterosexual Matrices of the Toilet

Interviewees notice how gender recognitions in bathrooms are mediated by the logic of heterosexuality. The architecture, design, and codes of conduct mandated in public washrooms secure heterosexual matrices by gendering excretion. How one urinates is not only a sign of gender but an indication of one's sexual orientation. There is a curious and none-too-subtle association between excretion and sex. As Neil, who is queer and transmasculine, notes,

> [That] you need to go pee has nothing to do with what you do in a bedroom, but for some reason, it has everything to do with what you do in the bedroom. Because you enter a female or male space, it's assumed that ... you like the person in the next [room] and when you contest [gender] categor[ies], then automatically your sexual orientation comes into question ... Straight away it raises a whole issue with homophobia; it's weird how they are interrelated in a bathroom. (Emphasis added)

Presumptions about an essential relation between gender, sexuality, and excretion are made in bathrooms. When one's gender is seen to depart from culturally authorized masculine or feminine scripts, heterosexuality is often called into question. There is also, as Neil intimates, a means through which urination is sexualized. JB, who is a genderqueer butch lesbian, notices a similar association between sexuality and elimination: 'I think that we have ... cultivated this society that thinks that washrooms are about your [sexuality] ... about sex somehow ... anything that has anything to do with your genitals is somehow ... sexual.' As Thomas Crown, who is of French and Belgian ancestry, a non-trans genderqueer man, and gay, says, 'I think that washrooms are simply a microcosm of how we understand gender and sex and sexuality.'

Equations between gender, sexuality, and excretion are so common in Canadian and American culture that they are often unquestioned. Sarah, who is transgenderist, remarks that 'For most people, gender just is. "I'm a guy, I like girls. I gotta go to the bathroom." ' Gender normativity is bound to heterosexuality in much the same way as it is bound to toiletry habitus. Urinary practices are not unrelated to the cultural politics and sociality of sex. Heterosexual matrices must be understood to orchestrate what people do in bathrooms. Toiletry practices encompass an entire dimension of sexuality that has been largely ignored, except within the circle of gay male public sex cultures and academic and community-based writing on tearooms, cottages, and bathhouses.[5] The gendering of public toilets is so beholden to reigning ideas about sexuality, and about heterosexuality in particular, that it almost escapes notice. Those who are queer and/or trans, those who have their gender called into question in bathroom spaces, are most likely to question the heterosexist and cissexist design of the toilet.

For example, Michael Silverman, the lawyer representing Helena Stone, who was arrested for using the 'wrong' bathroom at Grand Central Station in New York City in 2006 (see the introduction to this book, note 5), comments on the heterosexism informing gendered washroom designs:

> Bathrooms are designed ... [with] heteronormative standards ... they were designed because the sexes should be separated because of sexual attraction, and that doesn't count for gay people ... The whole thing is silly ... I never heard in public discourse anyone suggesting that there should be gay or straight bathrooms. That would be a hoot. I would love to see how that plays out. Can you imagine in the theatre when the line is long in the heterosexual male bathroom, and people would be like 'just use the gay one.' Yes, they are clearly designed with heterosexuals in mind.

One trans interviewee suggests that gendered toilet designs reveal heteronormative worries about genitalia.

> If you make sure that everybody is separated out according to their genitalia then, you know, you won't be exposed to strange genitalia, right? [Interviewee laughs.] Women can encounter other women's genitalia, that's okay; men can encounter other men's genitalia, that's okay, but my god you can't have the sexes encountering each other's genitalia ... Let's say that bathrooms were not gender segregated ... there's a urinal and there's a man standing at the urinal and a woman comes in to use the stall – *Oh my god*, there's a man standing there with his penis in his hand ... He's just peeing for god's sake ... I've never heard of anyone suffering any ill effects of being in ... [a mixed-gender] bathroom.

As an interviewee who is a non-trans queer femme explains,

> I think there's anxiety around heterosexuality that happens in bathrooms, too. So it's like, 'Hey, dyke, what are you doing in this bathroom?' And so

> gender policing also manifests itself as well as homophobia ... I feel like the bathroom is a super-heterosexual space, and that you need to perform that.

As Neil notes, 'I get weird looks [in washrooms] and even if I was perceived as feminine I'd get weird looks because I wasn't perceived as the right kind of feminine, which was straight.' Gender recognition is, as Neil says, often conflated with sexual orientation. Bryan, who identifies as transmasculine and advocates for gender-neutral bathrooms in the San Francisco/Bay area concludes that gender-segregated toilets are, in fact, mandated by heterosexism. He elaborates: 'A lot of that [segregation] does have to do with homophobia and that's a huge underlying cause ... This division, it's very heterocentric and homophobic, you know this idea that men and women have to use the bathroom in separate places.' David, a non-trans gay man, comments that

> The washroom is ... itself a bit heterocentric ... [The] assumption [is] ... that ... women who are presumably attracted to other men can go to their women's washrooms ... [and] powder their noses and fix their hair. Likewise, the men can retire to the men's room where the object[s] of their desire are not there and perform their bodily functions and straighten themselves up ... For a gay male, that is certainly not the case.

Because bathrooms are meant to be heteronormative spaces, it should not be surprising that there are latent (and sometimes manifest) worries about homo-sex in toilets. Butler's (1990) analytic of the heterosexual matrix maintains that gender stasis is purchased through refused homosexual (and incestuous) attachments. Gender is also scripted by what Lee Edelman (2004) calls reproductive futurity: an identification with white heteronormativity and its future, along with a concordant disidentification with the sodomite, who is future-negating, licentious, and unresponsive to reproductive and familial ethics. It should also be noted that the future is coded as white, and the incest prohibition, like the homosexual prohibition, is invested with a 'eugenic significance: the race must be protected from the results of consanguineous marriages ... there is nothing more common today than belief in the degeneracy of the children of an incestuous union' (Bataille 1957, 199). Today we may add that there is nothing more common than the perceived threat of the homosexual (as pederastic sodomite) or the trans subject as gender deviant to the sanctity of the family and its future.

Heteronormativity gives credence and value to ideas about progress, linearity, patriarchal succession, and the privatized white nuclear family and its health – all of which function to invest normative gender positions with a puritanical and familial importance.

The modern-day homosexual or the 'sinthomosexual,' who is, in Edelman's analytic, pleasure seeking and future negating in his or her disregard for convention, has unproductive sex and is therefore aligned with the death drive in the Symbolic register. His or her sex acts in the lavatory (itself a repository for the grotesque) are subject to prohibition. An obvious example of the heteronormative injunction on gay male public sex is seen in the removal of cubicle doors (but only in the 'men's' room).[6] The preoccupation with closet doors reveals worries about gay male sex.[7] For example, when asked about how municipalities and authorities police sex in toilets, one non-trans gay male interviewee in Toronto[8] said,

> They have been doing things like eliminating or locking open washroom doors so there is a loss of the privacy or warning that men can have when having sex in a washroom. I have seen doors on stalls removed which ... prevents any kind of action in there, but it also hampers someone's real use because they don't feel comfortable sitting on a toilet with the door wide open.[9]

He also notes that 'newer washrooms are usually designed without doors, so you get no warning [that someone may be approaching while sex is in progress] and someone can walk in at any particular time.' This interviewee explains that the best architectural designs are those that have 'blind corners, [with] the urinals ... way around a corner [where sex can take place] or [where] there are two squeaky doors that a person has to get through before they'll actually see anything.' Brant, who is of Scottish and French ancestry and a non-trans queer man, similarly focuses on the positioning of the door:

> Where the door is positioned is very important ... If I'm sitting there jerking some dude off at the urinal and the door opens, I know that I'd want to keep it under wraps for a couple seconds, and then start up when I know if the other guy is playing as well, or if you want to switch and go to another location or whatever ... But I'd say the common thing is the leeway; if you can hear the door open then you'd know someone is coming ... whereas if someone opens up the door and it's right onto the urinals,

you're pretty much fucked, because if you are jerking off a guy and you get someone watching that's not into it, you're going to end up in trouble.

Ivan, who is also a non-trans queer man, laments the new architectural impediments to sex in toilets:

> Old-school bathrooms ... [are] great for sex ... the door is noisy, and you have to go a long way around [a corner] before someone views the play area [by the urinals] ... What I've noticed now is you have these new-fangled bathrooms [that] don't have any doors at all ... so there are conscious decisions being made about preventing men from having sex in bathrooms through architectural design ... And I hate that. Because it [sex] is technically ... unauthorized activity in bathrooms. But ... heterosexism is an unauthorized practice in my books ... So ... how do you fight against something like that?

Not only are doors removed, but the metallic wall panel in cubicles is sometimes cut to reduce its screening function. Brant, who has travelled extensively in the United States, notes that stall doors in highway rest-stop bathrooms are becoming shorter.

> The distance from the ground to the top of the door was bigger and the height of the door was probably about four or five feet high. So if you are sitting down on a toilet you can look up and you can actually see someone ... They are eliminating the privacy that you can have in a public space, and it definitely served as a deterrent for me having sex in that bathroom.

The height of stall partitions and doors is not inconsequential when it comes to panoptic institutions. Architectural designs that give rise to disciplinary power cater to visual modes of surveillance. But as discussed in chapter 3, the panoptic gaze theorized by Foucault (1979) is splintered in the modern toilet. It gives rise to multiple lines of vision and points of interception. It is noteworthy that 'lighting' and 'photography' – both relating to vision – figure prominently in histories of entrapment and police surveillance in Toronto toilets (Maynard 1994). Undercover police have also been known to use ladders to peek over partition walls. The position 'from above' enabled officers to bask in what Christopher Lane (1999) would call a jouissance or 'ecstasy of bigotry' (266). The 'ecstasy of bigotry' is about entrapment, about a will to 'steal' another's (homosexual) 'pleasure by insisting that we renounce and sacrifice its pleasure' (ibid., 265). The ecstatic moment is transferential. It is also ensconced in a regime of disciplinary power that enables the law enforcer to 'see without being seen' (Foucault 1979, 171) by the one under observation. Insofar as the one doing the looking is presumed to do so on behalf of the law (and not for his or her own vicarious pleasure), the homophobic voyeur enjoys the sex he must and does refuse.[10]

But the homophobic voyeur is a product of prohibition, and his sexuality is not neutral with respect to homoerotic mirror circuits. Interviewees discuss how toilet pornography hung on walls (usually pictures of pin-up girls) aligns the masculine subject with the position of hetero-voyeur (even when he has homoerotic desires). The positioning of the masculine body at the urinal amid heterosexual porn (unofficial) and heterocorporate advertisements (official)[11] is indicative of a visual economy of hetero power operative in the toilet. Heterosexual body maps (illegitimate and legitimate) cover the walls. David explains:

> in front of urinals where you stand there'll be framed advertisements ... I've seen ... heterosexual-specific things up there, like pictures of the sunshine girls ... I see it quite often in decorations in men's washrooms that have or espouse a heterosexual idea, either a picture of a scantily clad woman, I've never seen male nudity, and other sorts of things that propagate the assumption that everybody who uses that washroom [is] not only male but heterosexual. That's one thing that bothers me the most. As a gay man, that has bothered me the most of all the other harassment I've received.

Callum, who is a trans guy, says, 'I feel like every time I sit down there's an ad right in front of me, and it's a very specific, male-oriented ad around hockey or beer or condoms ... or cars ... depending on if I'm ... [in] more financially privileged communities.'[12] Zoe, who is a non-trans woman, recalls a comment by a male friend that the toilet in his workplace is a 'very heterosexual, hyper masculine space ... [where] "gay" men are not welcome.' She also said that the toilet in his workspace was

> a space for heterosexual men ... [they] stood [during] break time [which] was [also] masturbation time as well. So, not only were the walls papered with porn, but there were porn magazines in all the stalls. So, it was understood that men went in to masturbate during break time.

There is, as Zoe notes, a means through which a braggadocio enactment of white, non-trans, masculine heterosexuality is almost expected of bathroom patrons.

Interviewees also observe a self-conscious preoccupation with male homoeroticism. The self-consciousness manifests itself in strange conventions governing the spatial configurations of urinals and the choreographies they inscribe. The floor space before the urinals is ridden with homophobic anxiety. Seo Cwen, who is trans and genderqueer, puts it succinctly: 'There's just a general fear amongst straight guys that the gays are going to use the urinals as an opportunity to check you out. There is also a fear amongst "men" identifying as straight that they will be perceived to be gay.'

David describes the fear-based protocol: 'You pick the urinals to optimize distance, to spread yourself out the most. That's the general rule; so if there's three or five available you'll probably pick the furthest one from the one that is currently being used.' Rachel, who is transgender and lesbian, agrees:

> There's rules about which urinal you pick ... you go to the furthest one from the door, and the next person takes the one closest to the door. You never go stand beside another guy if there's an open urinal ... [there are] very strict rules about it ... it's an unspoken code of conduct.

Ivan also says: 'I think people are very guarded and circumspect about where they stand.' Brant confirms that there are homophobic rules governing personal space before the urinal:

> if there are three urinals and there's a guy on the end, you are not supposed to stand in the middle because it makes people feel uncomfortable, because it's getting too much in their private space. But if there's two people on [either] ... end, it's an acceptable thing to go in the middle ... I like breaking bathroom etiquettes [governing urinals and how close you can stand to other men]. If a guy is on the end, I'll go to the one in the middle on purpose. You can see them squirm, and I think that's kind of funny.

Ivan comments on how urinal designs enact a regulatory function:

> I find myself ... monitoring myself, regulating my own body because of the heterosexist nature of men pissing in front of each other with their

cocks out, standing side to side. Like when you think about it, it's kind of strange, a whole bunch of men standing next to each other with their cocks out. And so what happens because of the heterosexual nature of social space is that I end up being very hyper-aware of my head movement.

Ivan also notes that the heterosexist designs and protocols of the 'men's' room produce a curious 'split':

> I have this split existence, where I've had very sexualized time in bathrooms, but then if I'm in a situation that's obviously not sexual, I monitor myself, so that I'm not looking at the guy's dick, and looking straight ahead, and I'm not standing too far back from the urinal ... I'm not indicating anything ... You monitor ... what you say when you're pissing, because you know that other straight men are around, and they're probably feeling more vulnerable. So, that increases the risk and the danger.

Despite the threat of homophobic assault, many non-trans gay male interviewees have public sex in toilets.

Interviewees do recognize the importance of public sex to queer cultures, but regret its relative exclusivity to gay, white, able-bodied non-trans men. Jersey Star, who is bisexual and polyamorous, says,

> I wish that there was a kind of toilet culture for women, like that kind of cruising space. I think it would be great. You know if you are a ... fourteen-year-old baby dyke wanting to get a little taste of what lesbian sex is ... you could just go down to a public toilet in a park and find out ... I think that would be fucking awesome. But it's not, unfortunately; for men, for gay men, it is.

Crystal, a Chinese-Canadian living in Toronto, who is a non-trans genderqueer woman and lesbian, offers her thoughts about gay male public sex cultures in Toronto:

> It's really brave and really exciting and I think it's really great that they've carved these spaces out for themselves. I think again it's related to patriarchy and sexism that men ... feel more entitlement in these public spaces than women do ... There is a culture that develops ... rules ... norms and [this] makes it easier [for gay, white men] ... And that same culture just doesn't exist for women. There is almost no [venue] ... there is [only] a [Toronto women's bathhouse] three times a year.

Gay male public sex cultures in bathrooms are not only exclusive to cis-sexual men but racialized. For example, one white gay male interviewee was troubled by racism he saw directed towards Asian men. This interviewee focuses on cruising styles in Toronto and the feminization of Asian men in mainstream gay male cultures.

> Certain cultures or 'races' ... are discriminated against and not approached [by white men], so ... [they have] to be more assertive. It tends to be obvious [that those racialized as non-white are] ... not getting enough [sex] on their own so they are pushing to get any kind of action ... In my experience it tends to be Asians. Anyone [presumed to have] a smaller size penis or they may not look like they are as masculine as they could be ... they tend to be more assertive, more obvious and overt [thereby increasing the risk of being caught by authorities or sex-negative patrons].

Brant, similarly, notes that,

> In terms of cruising for sex ... a lot of cases depending on your area, like in Abbotsford [British Columbia] a lot of people seek out white people if they are white ... You just see how cruising activities get racialized as well, and you can see cultural differences in the way that people cruise. Because it doesn't meet the white standard, some people get shunned or made fun of or people won't have sex with them based on racial or visible racial characteristics ... a lot of Asian people get really stigmatized.

Brant also remarks upon how cruising styles in Canada and the United States are 'predominantly defined ... by those who ... [are] white ... So people who are cruising have to conform to the white code or else they get marginalized in the cruising space.' He also notes how Asian men are constructed as 'feminine,' 'passive,' and as having 'small' penises.[13] In his discussion of Asian-American masculinities, David Eng (2001) reminds us that, 'If the symbolic order is always also a set of racializing norms, it becomes impossible to speak of the heterosexual matrix apart from racial distinctions' (141). Dominant white masculinities are dependent upon an idea not just about sexual difference but about racial difference as well. By constituting Asian men as effeminate or as having smaller penises, gay white men are able to assume a more rigidly masculine subject position in queer economies of power. According to Eng, 'a white male is placed into the position of being more masculine through his disavowal of the Asian penis' (ibid., 151). There is a related process through which black men are racialized by white men as hypersexual and virile. In white heteronormative economies of desire, the sexuality of the black man 'threatens the unity of the white male ego by placing him in the position of being less masculine, thereby endangering the structural distinction between him and the white woman' (ibid.). Even though heteronormativity may no longer operate as a primary signifying system in gay male public sex cultures, race does figure prominently in the way gay men lobby for erotic capital.

Eng further suggests that the 'power of symbolic norms of heterosexuality and whiteness functions largely through the tacit veiling of their collusionary ideals' (2001, 161). Heterosexuality and whiteness work in tandem.[14] Both enable those who can pass as white, cissexual, and heterosexual men to attain a privileged position in the signifying chain. The way in which Asian masculinities are feminized by presumptions made about their anatomical shortcomings (phallic impotence) is not unlike the process through which 'women' and some trans men (who use the stall as opposed to the urinal) are positioned as castrated subjects. Insofar as cissexual women are positioned as objects for heterosexual male voyeuristic pleasure, we may assume that, like Asian and/or recognizably trans men, they are used to conceal the homosexuality that is disallowed in masculine public spaces. White non-trans heterosexual men often refuse to be aligned with castration (coded as feminine), in part by insisting that Asian men (not themselves) are not so well endowed.[15] By refusing femininity, they are also, by implication, refusing what is assumed to be a passive homosexual position.

It is perhaps not surprising that prohibitions against homosexuality are most rigorously enforced at the urinal, where it is not only the penis but the anus (coded as feminine and as Asian)[16] that is susceptible to view. While the anus remains covered by clothing, it is lined up for show from behind. This is what Lee Edelman calls '(be)hindsight': the 'supposition or imagining of the sodomitical scene' (1991, 101), which can, as he suggests, 'wound the non-homosexually identified spectator who is positioned to observe it' (102). The spectacle of the anus, overdetermined by sodomy and racialized as 'dark,' threatens to reverse hetero- and homosexual positions. One is disoriented by the backside because heterosexuality demands attention to the front. Because all bodies – male and female alike – have anuses, they confuse and do not ensure sexual difference. (Be)hindsight obscures genital difference, whereas the front enables it. We are all 'feminine,' or, perhaps, homosexually predisposed, from behind. Edelman notes that this disorientation of positionality is bound up with the danger historically associated in Euro-American culture with the spectacle or representation of the sodomitical scene between men, and that this can be demonstrated by attending to the ways in which the logic of spatio-temporal positioning insistently marks our culture's framing of homosexual relations. (1991, 103–4)

There is, in Edelman's words, a 'phallocentric positional logic' (ibid., 104) at work. Attention to the anus or to the posterior confounds the 'stability or determinacy of linguistic or erotic positioning' (ibid., 105). Urinary positions are based upon repudiated pleasures, anal eroticism in particular:

> the male here must repudiate the pleasures of the anus because their fulfillment allegedly presupposes, and inflicts, the loss of 'wound' that serves as the very definition of femaleness. Thus the male who is terrorized into heterosexuality through his internalization of this determining narrative must embrace with all his narcissistic energy the phantom of a hierarchically inflected binarism always to be defended zealously. His anus, in turn, will be phobically charged as the site at which he traumatically confronts the possibility of becoming 'like his mother,' while the female genitalia will always be informed by their signifying relation to the anal eroticism he has had to disavow. (Ibid., 106)

The spectacle, or fantasy, of sodomy in the toilet is not only gendered but spatially mediated. The heterosexual logic of spatio-temporal positioning is well noted by interviewees who comment upon how one must take up (or refuse) space before the urinal. Urinary positions keep cissexism and heterosexism intact, as they can also undo spatio-temporal frames in which they are secured. By the same token, defecation is closeted because it sensates an orifice that is less susceptible to sexual difference. 'For the public insistence on the visible organ in the open space of the urinal can never dispel the magnetizing pull of the dangers that are seated in that unseen space, that cavity concealed by the toilet stall door' (Edelman 1994, 169).

There also seems to be anxiety about how body parts, the penis and the anus in particular, may be subject to the homosexual or queer desires of others. Edelman (1994) suggests that the 'historical framing of the men's room as a focus for straight men's sexual anxieties condenses a variety of phobic responses to the interimplication of sphincteral relaxation and the popular notion of gay male sexuality as a yielding to weakness or a loss of control' (161). While the anus must be tightly closed, the phallus must be visible before the urinal. Prohibitions on looking paradoxically enhance (and draw attention to) the penis (as discussed in chapter 3). If virile masculinities depend upon the display of a penis (masculinized sexual organ or silicone substitute) to other men, there is a curious (although disavowed) homoeroticism at work. Rico, who is a non-trans gay man, elaborates on what he regards as a homophobic fear of exposure. The fear is less about the display of an organ, or body part, than it is about a disowned homosexual desire: 'In a heterosexual environment there's ... definitely more of a focus ... [on] not ... being aware of who's around you, especially in the men's room ... on fear of desire, and fear of exposure.' Thomas Crown further explains that 'You keep your eyes to yourself and keep things covered [in the men's room].' Speaking about dividers between urinals, Brant says that they are built to prevent men from seeing each other's penises: 'I think the dividers [signify] really big privacy issues for people so people won't do the sneak-a-peek. Maybe the dividers are there to deter public sex.'

Prohibitions in the 'men's' room are intricate and comprehensive. As Matt Houlbrook (2000) observes in the British context, 'Thou shalt not need to urinate too frequently,' 'Thou shalt not stand in the stall next to a man, when there are other stalls free,' 'Thou shalt not talk to another man,' 'Thou shalt not make eye contact or smile at any other man,' and 'Thou shalt not undo the buttons on thy fly any more than is absolutely necessary or appear to have thy person on display more than the court deems fitting' (59). Most of these homophobic codes of conduct were noted by North American interviewees in this study. Brant, for example, focuses on the rules governing speech and how he violates the codes of silence:

> Another thing is that you are not supposed to talk, and I talk all the time. I'll turn to the guy next to me and look at him and say, at his face, not to sneak a peek, 'Hey, how are you doing? What's going on?' Sometimes it makes people really uncomfortable, too, but at the same time, that's who I am, and part of me wants to see if they'll get uncomfortable and part of me wants to see if they respond.

The 'men's' room is so thoroughly saturated by a negated homosexuality that it is almost overdetermined by the spectre of homoeroticism.

Zac, who is white, of English heritage, and a non-trans gay male living in Toronto, elaborates on his feeling of discomfort in the toilet:

> When I'm at the urinal sometimes I feel awkward because I don't want people to feel like I'm checking them out ... I make sure ... not to fall into any of the stereotypes that a gay man might fall into sharing a bathroom with a whole bunch of heterosexual people. Meaning, I'm not staring down [at] other people's penises while they are urinating.

David reflects upon how homophobic stereotypes about predatory gay men affect his comfort in toilets:

> If I were perceived to be a gay male in a men's washroom in a non–queer-positive sort of space ... other men might [be] uncomfortable or sort of make assumptions about how I would like to violate their privacy ... I definitely feel [a] greater level of comfort in a ... gay bar, so, for example, I [would] stand next to somebody at a urinal and pee ... I would be more comfortable doing so at a gay bar.

Ivan says that by simply using the 'men's' room he feels as if he is 'buying into the heteronormative systems of power ... because ... it's almost like everyone there is assumed to be straight.' In order to disrupt the heterosexism of the bathroom, Ivan doesn't refrain from touching his partner while at the urinal:

> Sometimes I'm there with people I'm dating ... [or my partner who was] waiting [in line with me] for the urinal ... He was standing in front of me. And I got closer to him, and I was touching him a little bit, and I was kind of leaning over him, into his space ... [because] I don't just want to stand here like isolated heterosexual units. I want to disrupt that.

Heterosexual Writing on the Bathroom Wall

LGBTI interviewees frequently see homophobic graffiti[17] in bathroom stalls. John, who is white and living in New York City, a non-trans gay male, reports seeing 'the word "fags," "death to fags" written on the wall.' As Charlie, who is Jewish and living in San Francisco, a non-trans genderqueer male, bisexual and polyamorous, notes, 'It depends on where I am, but I have seen plenty of graffiti about so and so is a fag, such and such is a homo or kill all fags or stuff like that.'

Homophobic etchings are, however, ambivalent. They reveal something about desire and taboo. Graffiti may be read as a secretive or confessional testament to what may otherwise have been foreclosed by our manifest sexual orientations and identities. Carol Queen, who is of northern-European ancestry, living in San Francisco, author of *Real Live Nude Girl: Chronicles of Sex-Positive Culture* (1997, 2002), a sex educator, a non-trans female, bisexual, and polyamorous, acknowledges the homophobia central to toilet graffiti, but also considers how it contains an element of the repressed:

> There's an element ... [a] sort of secret ... wish about it too. Where someone who doesn't necessarily have space in their real life to act out what they're acting out ... feels that ... [the wall is] a place where it's appropriate ... because it's so anonymous, and because other people have done it, already. The function of the graffiti in the restroom being ... [an informal] space ... [where] discourse has started, people who wouldn't otherwise [put] graffiti anywhere, feel perfectly free to add on their comment, because it's some level of public bulletin board.

Graffiti is, as Carol Queen suggests, multilayered and dialogic. The written exchanges reveal intense erotic predilections and distastes, love and hate, reverence and loathing. The ambivalence and capriciousness of graffiti are well illustrated by polarized dialogues about queer sexual practices. Madison, who is queer, comments on how homophobic graffiti is often a response to positive affirmations of same-sex desire: 'When homophobia erupts in graffiti, it usually comes out of, like, "Sally loves Jane." Like, a positive professing one's love for somebody of the same or not-right sex ... it turn[s] into some argument ... [in which someone will write] "That's gross." ' The bathroom wall is, as many queer interviewees note, a contested terrain. Carol Queen suggests that 'it's a place where the [dominant] culture can act out these sources of anxiety [around sexuality] that are no longer quite as easily acted out in other contexts ... It's very polarized.' The polarization is often sexually explicit. Zoe, a non-trans woman, describes the dynamics of the contested terrain of the bathroom wall at her Ontario university as follows:

> The public toilet in the Women's Studies area was one of the few places where you found queer-positive graffiti, and that created a certain sense of community in that toilet ... [There were] positive images of women's

bodies, and talking positively about lesbian sex, and queer sex ... that was really great ... [Also] the fifth-floor bathroom of the library at my old university was ... a hotspot of bathroom sex ... [according to] men's anecdotes of that bathroom ... people would write the call number of a certain book on the stall wall and then that would become a meeting place for people to go and hook up ... for sex. But ... once that bathroom became known as a hot spot for gay sex ... homophobes ... appropriated that use of the wall ... [to] write call numbers [entrapping men in pursuit of public sex]. And then I heard of people getting beaten up after having gone to that call number. So it was really tough because in some senses the community was formed [in a good way] and then [it] was infiltrated in a bad way.

Ivan, commenting on competition between gay-positive proclamations of desire and homophobic rebuttals, observes that

The bathroom walls seem to be a space for ... the unconscious ... layered with stuff ... there's like openly queer stuff like: 'I want head.' Or, 'I like to suck cock' ... There's queer stuff, sometimes there's drawings of guys fucking and giving head, sometimes there's drawings of heterosexual sex. But I've also seen more gay-affirmative graffiti competing with the homophobic [in recent years].

Much of the homophobia is sexually graphic. According to Rico, there are 'a lot of ... sexual drawings ... of genitalia, bodies, male, female ... Things drawn way out of proportion, depictions of acts, sexual acts, phone numbers.' The pictures are often of sodomy, cunnilingus, or same-sex acts.[18]

Graffiti on the bathroom wall is not only homophobic but misogynist.[19] This is well illustrated in the following story told by Diane, who is lesbian, about graffiti in the law school toilets at her university:

There would be graffiti that talked about them [female law students] all being lesbians ... [you would see] gross pictures of ... professors and some of it [was] about students, about what their bodies ... look like. [The images] would often be masculinizing their bodies, very hairy and ... describing a sex act they liked to engage in ... [accusing] any of the men that supported feminists as being gay and ... so and so sucks dick ... We went in and blacked it out ... [but] the next day it was there again and we finally got them to start painting the washrooms every night.

Even when female law students are clearly feminine and heterosexual, they are masculinized and subject to misogynist and homophobic slurs as punishment.

Sexual references are central to the shaming process. As JB jokes, graffiti 'usually has a sexual link ... nobody's ever written on the bathroom wall ... "JB has excellent grades" ... [laughter] I've never read that, never! It's always ... "Johnny's a fag", or [someone is] hot [or it contains] ... some sort of sexual threat.' Isaac, who is a trans man, bisexual and polyamorous, remarks: 'What I see is usually Joe sucks cock, Jon is a fag, Sara is a slut.' Lisa, who is a white, non-trans femme lesbian living in Kingston, Ontario, comments on the misogyny central to much of the graffiti she sees: 'Usually I'll see something with a woman's name and "so and so is a slut" or a "ho" or "is a dyke." They are usually sexual slurs [about] somebody's sexual behaviour.' Eric Prete, who is queer and polyamorous, agrees: 'The graffiti in high school [toilets] is ... sexual. It's, like, who's a whore, and who's gay.' Madison confirms that 'if you hate somebody [and she is female], you write their phone number [on the wall in the 'men's' room, in front of the urinal] ... Like it's this thing that you do in bathrooms ... you write this graffiti that has to do with sex ... For a good time call ...' Pam, who is a white American (living in Kingston, Ontario) masculine/butch woman (who also identifies as transgenderist and genderqueer) and lesbian, insists that graffiti is 'all about sex. It's, you know, negative judgment of sex or positive pronouncement of love. It's mostly about sex.'

Interviewees notice how sexualized – usually homophobic – writings are also racist.[20] Rohan, who uses the 'women's' room, pinpoints the way homophobic graffiti intersect with racism: 'It's often homophobic, occasionally racist, and occasionally anti-Semitic. But it seems the majority of the time it's homophobic.' This interviewee also notes that 'a lot of graffiti is very bigoted ... very sexist; it's homophobic, it's sizeist, it's racist, it's anti-Semitic, I mean it's transphobic. I've seen the entire gamut.' As another interviewee explains, focusing on the American South, 'when you [go] to the restroom you always see, "die nigger die," "KKK rules," "fags are bad" ... hate types of writings ... I've seen a lot of gang tags.'

When asked specifically about transphobia, many interviewees suggest that transphobic sentiments are subsumed into homophobic etchings. In Charlie's experience,

> I don't think I have seen a lot of trans stuff but I think there isn't also a single word that covers trans folk in the way that fag or queer conveys, so it doesn't lend itself to graffiti in the same way. I don't think I have ever seen anything like 'Kill all genderqueer and gender-deviant folks.'

This view is echoed by Syd, who is of Japanese and western-European heritage, living in the San Francisco/Bay area, identifying as a boi/fag and drag-queen bitch, queer and polyamorous:

> I've seen more homophobic graffiti ... like ... 'So-and-so's a Faggot,' or 'so-and-so sucks cock,' or something like that ... people who are transphobic ... often lack the accurate language to be [trans]phobic [laughter] ... so ... most of it is just that kind of ... colloquial use of, like, gay or faggot.

As Syd and Charlie suggest, transphobic hate is often conflated with homophobic writings and imagery. The conflation may be structured by a lack of available terminology to shame and condemn trans people as a population distinct from those who are designated lesbian, gay, or bisexual. It might also be that transphobic intolerance is signified by homophobic rhetoric. Foucault tells us that discourse sets parameters on what is thinkable, what is designated problematic, and how people are subsumed into various regimes of power. The conceptual and linguistic slippage between homophobic and transphobic graffiti indicates a culturally inscribed conflation between trans subjectivity and homosexuality. Insofar as homophobia is a 'prejudice of categorization' (Young-Bruehl 1996, 150) and those recognized to be 'gay' upset heterosexuality's rigidly choreographed gender scripts, it must be a phobia incited by gender as well as sexual transgressions.[21]

Mapping Homophobic Panic in the Lavatory

The interview data suggest that gender and sexually nonconforming people are seen to threaten cissexual women (as discussed in chapter 2), cissexual men who are heteronormative (with homosexual rape or seduction as discussed above), and children (to be discussed below). Each supposed victim is culturally coded as white, heterosexual (or 'innocent,' in the case of children), and a key player in the nuclear family ensemble. A closed and exclusive unit, dependent upon binary spatial divisions between private and public and upon the domestication of genders and sexualities at odds with its reproductive mandate, the white nuclear family positions itself as a nationalist point of reference.

The child is a focal point of angst because his or her body (coded as angelic and virginal) plays a central role in the elaboration of white, heteronormative futuristic fantasies. 'The Child ... marks the fetishistic fixation of heteronormativity: an erotically charged investment in the rigid sameness of identity that is central to the compulsory narrative of reproductive futurism' (21).[22] LGBTI folks are perceived to interfere with the future because they are often thought either to interrupt binary gender positions (masculine and feminine) – their essentialist and unchanging coordinates as evidenced at birth and confirmed by medical and national identification papers – mandated by entrenched cissexist body politics, or to sexualize the allegedly asexual (read heteronormative) space of the toilet where senses are to remain dormant. Queers in particular are seen to occupy the underside of the heteronormative fantasy. 'For the only queerness that queer sexualities could ever hope to signify would spring from their determined opposition to this underlying structure of the political – their opposition, that is, to the governing fantasy of achieving Symbolic closure through the marriage of identity to futurity in order to realize the social subject' (ibid., 13–14). The trouble with queer and/or trans people is that they are perceived to interfere with the child's future. They are thought to endanger the child by their supposedly unlawful (unfamilial) and predatory sexualities, along with their seemingly unnatural gender identities, thereby undoing or upsetting the normative identificatory spaces earmarked as the child's future living space.

Because the threat is symbolic, it finds justification in *fantasies* about the endangered child. Futuristic imaginations often focus upon the pederastic[23] sodomite in the toilet. As Seo Cwen observes, 'I know some people still believe all queers are child-molesters and perverts, and those perceptions make going to the bathroom unsafe [for LGBTI people].' Sarah comments on how parents watch her closely in the toilet. Roxie, who is a visible cultural minority, trans female, bisexual, and living in the San Francisco/Bay area, says emphatically that 'People [are] afraid of us. They look at us as child molesters and perverts ... and ... hookers ... [and this may cause] ... problems for somebody.' Rohan says that on more than one occasion 'people come into a bathroom with their child, see me and grab their child and run out.' Focusing on how she is seen as pedophilic and purse-snatching, another interviewee comments,

> women will grab their children, and hold on to their children and hold their purses a little tighter. They keep an eye on you just in case, you know, who knows what you're capable of ... 'It's different than me, kill it, kill it,' you know. People don't like difference ... it's bad.

Rohan hypothesizes that,

> Anybody who doesn't fit [into normative sex/gender systems] ... is a threat, and is sick or perverted or a predator or a paedophile or clearly up to no good. You couldn't simply be in here [the toilet] because you need to pee, you must be in here because you want to hit on me ... you must be here to recruit.

While LGBTI people do not pose a threat to children, they are disproportionately regarded as dangerous. Elisabeth Young-Bruehl (1996) observes that 'Allegations about acts directed at children have always been a feature of obsessional prejudice ... the homophobias ... [in particular are] laced with charges that people (particularly males) who engage in homosexual activity are corrupters of children' (447). In fact, homosexuality and paedophilia are often conflated in homophobic and transphobic imaginations. The equation is irrational given the fact that most child sexual abuse is committed in the family home by heterosexual men who are known to the child, not by gay and/or trans folk in public space.

Those who have adult consensual sex in the 'men's' toilet are very likely to be seen as threats to children. Because gay men are regularly presumed to be sexually predacious and paedophilic, they are hyperconcerned about being so constituted and vigilant in their regard for youth. Non-trans gay men interviewed for this study go to great lengths to avoid being positioned as threats to children. One non-trans gay male interviewee, who regularly has sex in toilets, notes that he is always watching for children: 'There's always the chance ... someone will come in. It's something I am always on the look-out for because I don't want to be caught. It's not something I want to expose children or young boys to, any kind of sexual activity before they are ready to handle it.' Jay, who is genderqueer and gay, confesses that 'The one time that I was going to have sex in a public bathroom little kids walked in as we were contemplating it so we were, like, no we better not do this 'cause we were in Central Park.' Ivan notices that the late twentieth century saw a

> general increase of fear around protecting children from strangers and malicious people ... parents are [now] much more ... [likely to] monitor their children more when they go in the bathrooms. And I've been able to

notice because ... I'm queer and I'm in the bathroom because I'm wanting to have sex, and then a child comes in. Do you know what that does to someone? Can you imagine the associations? And then I notice parents are much more vigilant with their children in bathrooms ... But now, parents are so paranoid of public space, strangers, paedophiles, you name it.

Ivan also notes how worries about children provide a rationale for the policing of gay male public sex. Childhood sexual danger narratives[24] are, as he suggests, used to criminalize gay male sex rather than to protect those under the legal age of consent:

I realize how much of a live wire ... [the bathroom] is, because if you put queer men having sex in bathrooms, and you have a child element, it's an explosive mix. What if the kid walks in and sees you sucking someone's [cock] ... you know how people tend to problematize social practices because of their so-called deleterious effects on children. So this is like plutonium here, you have gay men using these bathrooms in unauthorized ways, then you have this discourse of threatening to children ... The danger of children would be cited all the time as a justification, a legitimization to clamp down and police such activity, and to regulate it even more, because a child might be harmed.

One non-trans gay male interviewee who has sex in public bathrooms is very conscious of how concerns about child welfare can turn violent: 'If ... you start cruising a man and it turns out that he is straight and his son is just in the toilet stall one over or in the urinal one over, sort of thing ... then he goes ballistic and freaks out on you and punches you or stabs you or shoots you.'

The anxieties of homophobic men often find expression in irrational worries about sodomitical seduction. Gay male interviewees regularly confront panic about what Lee Edelman calls the 'contagious disturbance of positional logic' (1991, 110) afforded by the sodomitical scene. Homophobic and cissexual men feel their 'sexual authority challenged by a sight that imposes upon the male a disturbing "consciousness of his own attractions" and thus an awareness of his susceptibility to being taken as a potential sexual object instead of an active sexual subject' (ibid., 109). The fear and fantasy of seduction is irrational and hyperbolic. For example, Ivan reflects upon how homophobic narratives construct gay men as preying upon straight men:

> Homophobic discourse would say that gay men are practically assaulting straight men in bathrooms. That's so much fucking bull ... gay men are so vigilant around [the] sex they have in bathrooms because of the stigma attached to it, because of homophobia, because of the threat of violence, the threat of arrest, the threat of social opprobrium, all kinds of things. The sex becomes furtive, and sometimes really dissatisfying. So, no, the sex that happens in bathrooms is so unobtrusive to the heteronormative space.

Seo Cwen, who is trans, confirms this:

> people think ... [that] if you go into the washroom on the third floor in the [Toronto] Eaton Centre[25] that you are going to get pounced upon [by gay men] and it's a total misunderstanding about cruising and it's unfortunate that there is a real lack of delineation between that type of [same-sex] activity and predatory sex.

He further elaborates upon the intersections between gender misreadings and heterosexual panic:

> Well, going into the men's washroom looking as I do, chances are that they'd either think I was a girl who'd 'wandered into the wrong one' and want to get me out of there, or, they'd see me as a 'cross-dressing fag looking to seduce gay men.' I haven't had any experiences with this, that's just the vibe I tend to get.

Roxie, who is a trans woman, offers insightful comments about how transphobic hate is sometimes driven by disavowed homoerotic desires.

> I think ... there's this fear that in a bathroom a line gets crossed. And I don't know, maybe if a guy is perceived as looking at an FTM, maybe he ... [thinks] there's something wrong with him. Maybe there's an inner fear that if I look at somebody who, like, used to be a female, you might have a little attraction to them, in your own mind you're saying 'Oh my God, what's happening to me?' And maybe you doubt your own sexuality and you get angry and you take that anger out on the person. Like, how dare they do that to me!

Sarah notices the fine line between male homophobic violence and disavowed desire.

> There's ... times when people have come to me ... and my partner, my lover, whoever I'm with looking for a fight, thinking that we're fags. And then as soon as they hear my voice they're, 'Oh my god, I'm so sorry ... you're a girl.' So now all of a sudden, they're no longer interested in kicking our ass, but they wanna fuck us. They're, like, 'Oh my god, lesbians, right on.' And they get all into it. And I'm, like, 'Five minutes ago, you were ready to fucking kill me, cuz you thought that I was a guy.'

Policing Sex in Toilets

> Arrested primarily in urinals, the homosexual was constructed in the image of that place ... [There is] the connection between the dirt and defecation of the lavatory and the homosexual. (Houlbrook 2000, 62)

> But does civilization need to cover its ass? (Berlant and Warner 2002, 196)

Perhaps the most obvious example of homophobic panic is to be seen in the alacrity with which police entrap and arrest gay men in toilets. As noted in earlier chapters, heterosexuality does not involve a wholesale refusal of homosexuality. Heterosexual panic is sometimes about a need to perform homophobia to satisfy a desire one has disowned in order to identify with heterosexuality. As Butler (1997b) explains, 'prohibition becomes the displaced site of satisfaction for the "instinct" or desire that is prohibited, an occasion for the reliving of the instinct under the rubric of the condemning law' (117). The 'men's' room may be one of the most intense sites of homophobic surveillance in North American public spaces. The surveillance is not, however, predicated upon a wish to exterminate homosexual acts, despite self-conscious hetero-architectural designs advertising interdictions on and obstructions to gay male public sex. Surveillance is a back-handed (or, shall we say backsided?) attempt to satiate a desire otherwise refused.

> The prohibition does not seek the obliteration of prohibited desire; on the contrary, prohibition pursues the reproduction of prohibited desire and becomes itself intensified through the renunciations it effects. The afterlife of prohibited desire takes place through the prohibition itself, where the prohibition not only sustains, but is *sustained by*, the desire that it forces into renunciation. (Ibid., 117)

In other words, the heterosexually invested law enforcer (or homophobe) needs the sodomite in the water closet.

The injuries inflicted by refused homosexual desires are many. Men who have sex with men in public spaces are subject to intense police surveillance and criminal prosecution – despite the consensual nature of most gay male public sex acts. Ivan, for example, emphasizes the homophobia driving the policing of public sex. He says that queer people 'have the wrath of the law down on us.' He also notes how hetero sex is had in broad daylight, whereas gay male sexual practices are far more likely to be hidden and discreet:

> Heterosexuals have sex in public too. They have sex in parks, on the beach, in cars. Sometimes in broad fucking daylight. I don't know about you, but I've never seen two men fucking in broad daylight, ever. I've seen heterosexual people fuck on the beach in broad daylight.

Kew, who is a polyamorous queer dyke, similarly notes how homosexual sex is singled out unfairly as a security issue:

> I don't go to many straight clubs but I've never seen people monitoring bathrooms in straight clubs. But in queer men or women spaces, there's usually people monitoring the bathrooms to make sure people don't go into the stalls together and hold up the line. So it's one person per bathroom. I've never seen that in straight spaces.

Liam, a white, non-trans gay male who was arrested and given a conditional discharge for public indecency, observes that public toilets in Toronto malls and in the subways are heavily patrolled by security guards:

> At the Toronto Eaton Centre there certainly are security guards who are patrolling. Around the time that I was arrested ... and I should have put one and one together, but ironically on CFTO news they had mentioned that the TTC [Toronto Transit Commission] had just hired undercover cops for that very purpose to monitor the ... washrooms [for gay male public sex].

Jay also notices how bathrooms in department stores are frequently checked: 'If it's a department store there's an hourly check. If it's known for being frequented by homosexuals, it is a definite situation when there will be more policing.'

Due to the heightened possibility of arrest, gay men take extreme precautions. For instance, Ivan emphasizes that he is

> very careful about every move I make [while in pursuit of sex], every move. I need to know the full lay of the land. Who is in there [is important]. I need to know they're queer. If I don't know something about someone, I will not deal with them [for fear of entrapment], until I know for sure, they're queer, and they're here doing the same thing. I'm extremely careful.

Tall guy, who is a white, non-trans gay male, takes similar precautions:

> I always listen for somebody to come in [the toilet]. I personally would never do anything in a washroom if you don't have any lead time to when the door opens, zip up get back to the urinal and cover up ... I always wait to be very sure that the person I might be with [sexually] is actually interested rather than a security guard or a cop looking to entrap.

Despite discretion and vigilant circumspection, Tall guy was caught and charged with public indecency. He offers the following account of the arrest:

> I've been charged with ... performing indecent acts in a public washroom ... The situation was that the person ... who I didn't know was a security guy was watching in through a stall. I was in the stall at the time. When he didn't seem particularly interested and I was playing with a guy at the urinals ... the man came up and said he was security and took us out to arrest us and have the police charge us ... It was his purpose for being there. It wasn't necessarily entrapment or being set up because he didn't do anything to entice any kind of behaviour but that seems to be the sole reason for him being in that washroom at that time.

Skyler, who is of British-Polish-Jewish ancestry, a non-trans gay man who works as a model, erotic dancer, and escort, once narrowly escaped being charged with public indecency. He recounts an episode in a toilet where a police officer walked in on him:

> I thought we were being smart about it [because the stall he was in was private and had a heavy door] and when the door opened it ... was a

police officer ... And *thank god* ... we had concealed ourselves so there was no actual public display of nudity and that is why we were let go. He said, 'I could charge you with an indecent act and it would be a $150 fine and in the end it's going to be a waste of my time and yours because I'll have to go to the law courts and you will too. This bathroom is not noise friendly – people heard you and it's not appropriate. If I had opened this door and you guys had been showing genitalia I would have arrested you.'

Ivan suggests that policing is very much about the production of distinctions between private and public. He also notes that there are different degrees of surveillance, depending upon the locale. Venues frequented by parents and their children are perhaps the most heavily patrolled. Venues predominantly inhabited by adults are less likely to be rigidly patrolled. Ivan elaborates:

I think maybe university bathrooms are a little different than, say, Maple Leaf Gardens. So ... yeah, it's context dependent ... playing [public sex] in bathrooms is generally safer in university bathrooms than at a mall ... like the Eaton Centre ... [where] there's more surveillance and policing.

The idea that sex should be private, monogamous, and confined to the marriage (or civil-rights-union) bed is hotly contested by queer people. For example, Brant adopts a gay and sex-positive position on sex in public:

Sex in public spaces in general ... is important, whether it's a bathroom or park or video arcade or a parking lot. I think sex in general is very important. I think it really raises questions ... [about] what is public space and what is private space and I like this sort of blurring of that boundary, because sex is ... intimate and personal and [dominant beliefs are that] it should be kept to the confines of the bedroom. I don't think it necessarily should be kept in the confine of the bedroom ... I'd like to see it going out in public spaces ... It's about space for me, and it's about ... confusing the public and the private, and I think it's a form of political activism even though many people might not recognize it as that ... I really think it ... [pushes] people ... to check their assumptions of what makes something private and public. And ... sex in public is illegal ... But people still engage in it, so I think it is political activism ... because of what it does to people's conception of space.

Ivan, who expresses frustration with heteronormative allocations of space, consciously eroticizes the prohibitions:

> There's a thrill that comes from cruising unauthorized spaces ... In the park, and in the bathroom ... because it's like saying a big 'fuck you' to the system, the heteronormative system.

Both interviewees regard queer public sex as political. By quarantining sex – confining it to the home – people can insist upon the asexuality of public spaces while at the same time imposing a heterosexist configuration upon those same spaces. The absurdity of the claim is noticed by Rachel. She comments upon how investments in private, asexual spaces, are heteronormative:

> It's heterosexuality, it's heterosexism, because ... you're doing very private things ... [with] your very private parts and [people typically] don't want ... [bathrooms] to be sexual spaces ... otherwise men wouldn't be able to pee standing next to each other if there wasn't this supposed heterosexuality. So, it's based on the idea that these are desexualized spaces where you just go and ... do your business.

The so-called desexualized space is designed to enable a cissexual heteromasculinity to forget that it is predicated upon queer divestments. The asexuality of the space does not involve a wholesale repudiation of the sexual. Heterosexualities founded upon a disavowal of homosexualities are allowed. There is a peculiar insistence upon the asexuality of spaces that are overdetermined by heterosexual self-identifications, their need to formalize and ritualize the repudiation of homosexual object choice. Homosexuality can be said to enable heterosexuality by its social status as a refused orientation. The supposed asexuality of the toilet is a closet for heteromasculine disavowals in the arena of the sexual.

But this refusal is curiously generative. The homosexual prohibition on sex ignites desire. Social taboos produce preoccupations with the illicit, a paradox incisively theorized by Georges Bataille (1957): 'The taboo would forbid the transgression but the fascination compels it' (68). He further suggests that the pleasure in taboo is 'mingled with mystery, suggestive of the taboo that fashions the pleasure at the same time as it condemns it' (107). The taboo against homosexuality (and against sex in public more generally) is felt to be exciting and furtive not only by the heterosexually identified subject (whose homosexuality is manifestly refused) but sometimes also by the homosexually identified subject, whose sexual identity is shaped by a conscious desire to flout the taboo. Raj, who is Pakistani, Muslim, living in Toronto, and a non-trans gay man, says that 'sex is almost enhanced in a place like the bathroom ... [it] is a place where you are more attracted sexually [to other bodies].' Pam also suggests that the 'kinds of things that can heighten excitement are things that are danger[ous], and so sex in a public place has the element of discovery and the possible danger associated with discovery, or the possible embarrassment associated with discovery.' Thomas Crown agrees: '[Sex in toilets] is dirty ... literally and metaphorically. It's raunchy and unsafe. It's illegal to have sex in a public situation ... So there's a level of drama and risk, and a lot of people like that ... There's just a different culture [of pleasure and excitement in public spaces] with gay men.' Claude, who identifies as queer, notes that 'our culture partly brings us up to think that sex is disgusting at some level. And I think that in a way the space of the toilet can be a kind of rebellion against that ... [it] is associated with disgust and sex ... and I'm going to turn it into something erotic for myself.'

Along with the eroticization of the prohibition on homosexuality, there is an intentional use and reclamation of space by queer people. For example, Carol Queen insists that public sex in toilets 'is fundamentally [about] queer people making space for themselves, and for their needs and sexual connections.' Kew believes that the 'bathroom provides a big opportunity for social things, so flirting, and cruising, and picking people up is a big part of that.' Haley, who is a non-trans queer femme, goes so far as to suggest that

> Having sex in bathrooms is part of queer culture ... Not that straight people don't have sex in bathrooms, but ... it's not the same, it's quite permissible in queer culture ... and ... if I was more actively sleeping with women ... I probably would have had sex in a bathroom. Because it's part of what it means to be queer.

Joe, who is white, of British ancestry, living in Montreal, and a non-trans gay male, notices how heterosexist rules about how one is to have sex and with whom are less likely to be enforced in queer venues: 'I think that in a gay establishment you are having gay men, lesbian women, trans-identified people, or even straight or bi people engaging in sexual activity, whereas in straight establishments you're probably more likely to find straight sexual activity going on.' Closely related to the break-up, or contravention, of the matrix is a space made available for those who are trans and/or gender non-conforming. Kew explains that 'the difference is ... [that] in queer spaces, gender is something that's actually more visible, questioned, and can be talked about ... I feel like the rules around gender and keeping everything pure and separated is not as functional there in those spaces.' Jay believes that bathrooms symbolize 'freedom to express and be and to enjoy yourself and enjoy all the stuff that goes hand in hand with having dirty hot [queer] sex.' Rico calls the toilet 'the icon of anonymous gay sex.'

Many interviewees who have sex in public spaces comment upon how they enjoy, and utilize, the instability between private and public in toilets. Joe describes sex in public as follows:

> I think there's this security in the fact that there is a high level of privacy and a high level of anonymity. So I think people can cruise there and have sexual relations and remain private about them and they can remain anonymous [even though they are in public venues].

Brant observes that 'norms of personal space change when you go into the washroom because even though it's a public washroom, it becomes sort of private.' Laura, Italian-Ukrainian, living in Toronto, a cisgendered (non-trans) female, and queer, explains that 'some people get turned on by sex in public ... it's ... slightly safer [in a bathroom because it's] less public ... you are doing it in a semi-public space ... some people find it hot or more enjoyable because it's dirty or it's ... unconventional.' Thomas Crown elaborates upon how gay male sex has, throughout history, been excluded from heterosexist spatial designs: 'Historically speaking, there are very few venues in which men have been able to cruise each other publicly and yet privately. And I think [the] bathroom [today] functions very easily that way, because it's public but it's somewhat private.'[26]

It is worth noting that excretion was not always private (Kitchin and Law 2001); nor was sex bound to the domestic sphere. As sexuality was increasingly confined to interior, domestic spaces over the course of the eighteenth century (Foucault 1978), excretion likewise became a more rigidly gendered and private activity. Just as the marriage bed was off limits to spectators, the modern latrine was increasingly quarantined. 'Privacy within these toilets was ensured by design guidelines and building codes, so that this public space was frequently divided by screens, demarcated by internal doors and shielded from the gaze of others by disrupted sight lines' (Kitchin and Law 2001, 288).

Reflecting upon the prosecution of sodomy in male urinals throughout history, Jersey Star infers that the toilet is strangely immemorial:

'this idea of cruising in public toilets seems to be this thing of the past where gay men used to meet up, and it's not [like it was, but] it still is. I really like that, toilets sort of transcend time a bit.' Star seems to read the toilet as a queer memorial – a living history, of sorts. Alive and fecund, yet archaic and germicidal, the bathroom is built upon the co-ordinates between prohibition and entrapment, desire and what we might call the emergence of queer counter-publics (Warner 2002). In his discussion of Foucault, Leo Bersani (2001) says,

> Nothing, it would seem, is more difficult than to conceive, to elaborate, and to put into practice 'new ways of being together.' Foucault used this expression to define what he thought of as our most urgent ethical project, one in which gays, according to him, were destined to play a privileged role ... what disturbs people about homosexuality is not 'the sexual act itself,' but rather 'the homosexual mode of life,' which Foucault associated with 'the formation of new alliances and the tying together of unforeseen lines of force.' (351)

Queer sex in public is, thus, productive. The spectacle of the homosexual in the toilet transcends time, as Jersey Star intimates, and signifies new modes of love, affiliation, identification, and desire, as Foucault and Bersani (2001) agree.

Queer and/or trans people are perceived to upset what Judith Halberstam (2005) calls 'conventional forward moving narratives of birth, marriage, reproduction and death' (314). White reproductive fantasies of the future invoke nationalist sentiments where 'culture can be imagined as a sanitized space of sentimental feeling and immaculate behavior, a space of pure citizenship' (Berlant and Warner 2002, 189). Those invested in heteronormative futures render gender and sexually non-conforming people abject: out of time and place. They are here and not here in much the same way as they are subject to persecution, rendered invisible, or projected into a negative (or abject) space. Those who have sex in public toilets exploit the border between private and public. They forge a presence on the cusp. As Magni and Reddy (2007) suggest in their discussion of sex in a South African gay club toilet, the 'bathroom cubicle entails a locking out of the outside world, as well as a constant reminder of the potential for "locking up" or imprisonment' (234). In other words, the toilet stands in as a reminder of what is forbidden and punished in normative cultures. Those who have sex in public toilets exploit the border between private and public, the proverbial 'gay closet' and its generative potential (Sedgwick 1990). As described by Layal, who is queer, cissexual, and Arab,

> [The bathroom is a] liminal space where you … [can] negotiate the edge of what can be exposed and what can't. The toilet manages us in that way and so those boundaries can be … pushed … it invites you to play with those boundaries and I think because of that it's appealing for queers because that is the condition of what it means to be queer is you are having to always negotiate what of your perverse private life do you render public. The anxiety that we live with, in that tension, is something that can be played out in the bathroom.

As Layal suggests, the toilet is not uniformly private or public but both, a tertiary or intermediate zone where the boundaries governing what is and isn't 'perverse' can be 'pushed' or exploited.

On the cusp of the private and the public, sex in toilets affords new ways of being sexual and animating the senses. The division between the 'grotesque' (coded as queer) and the 'clean' or 'normal' (coded as white, cissexual, and heteronormative) is overturned by those who count the grotesque as erotic. Jay, for example, embraces the grotesque in his discussion of sex in toilets. 'It's dirty. It's lusty. It's exciting … I would equate it [the toilet] to the same way I feel about seventies porn. It's got a gritty kind of feel to it – a gritty kind of rawness to it. You can't get much more animalistic than sucking some guy off in the washroom – especially if that guy is straight. It happens.'

In her discussion of pornography, Laura Kipnis (1996) explains that the 'power of grossness is very simply its opposition to high culture and official culture, which feels the continual need to protect itself against the debasements of the low (the lower classes, low culture, the lower body)' (137). Because LGBTI people are read as indecent, perverse, and abnormal (however differently), they will often resignify, eroticize, and flout the heteronormative prohibitions on public sex.

Jersey Star elaborates:

> I think what interests me about toilets is that they are seen as dirty and they are seen as places where you do dirty things … that's why I really like having sex in bathrooms … because you are not supposed to … it … fascinates me in the way that … water sports [do] … because you are just so not allowed to have those particular bodily discharges involved in any kind of sexual play and so combining what people read as totally dirty

and disgusting things and putting that into sex I really like. And so shagging in public washrooms, which are supposed to be these dirty, bacteria-filled places ... I like turning that into a kind of sexualized and a sexual environment.

As does Neil:

If there are two men who identify as men having sex in a bathroom, then I guess it has a sense of subversive power because it is seen as a dirty space and that they are reclaiming this dirt as their own. Which I think is great. I don't see anything wrong with it. I think it is a very powerful sort of thing, because you don't expect to see that in bathrooms ... [it] is ... [rigidly] gendered ... and it's usually a heterosexual, single-biologically sexed place. And if you [are] queer and identify as queer in all senses of the word, whatever that may be, then you are reclaiming the space.

Interviewees articulate a conscious and deliberate use of the grotesque to animate desire. Queer and trans-positive spaces are forged in the place of prohibition.

To reclaim the abject zone of the toilet and that which is putrid to heteronormative sensibilities is a queer and trans-positive manoeuvre. Forging erotic ties on the cusp between the private and public is instrumental to the project of reclamation. 'Making a queer world has required the development of kinds of intimacy that bear no necessary relation to domestic space, to kinship, to the couple form, to property, or to the nation. These intimacies *do* bear a necessary relation to a counterpublic – an indefinitely accessible world conscious of its subordinate relation' (Berlant and Warner 2002, 199). What Berlant and Warner call 'border intimacies' (201) are generative because they upset the heteronormative order that governs the 'who,' 'where,' and 'what can be' of gender, sex, desire, and the body. Gender and sexuality must signify in excess of heteronormative and cissexist house rules. In the next, concluding chapter, I offer suggestions for how we may reconfigure the exclusionary plumb lines of the gendered lavatory.

Conclusion

Gender is all about plumbing.
                              (Savoy, who is a non-trans queer femme)

Modernity is split from the beginning into plumbable and unplumbable depths.
                              (Lahiji and Friedman 1997, 47)

The plumb line comes out as a device that permits sounding into an invisible (or barely visible, yet refractive) domain where that which is to be translated or transported resides ... The plumb line thus embodies this condition and refers to a virtual 'ground level' that separates two domains (the image's and the object's) only related by the mirroring effect of the screen. As in the figure of Narcissus, and also as in *Vertigo*, this is a position of yielding to the place, of paralysis rather than of perspectival control. If the window stands for the architectural element that permits controlled, perspectival definition of space by the observer, then the reflecting screen implies captivation by space, as in suspension, in an arrested blind fall.
                              (Costa 1997, 100)

The modern toilet engineers a truth about the body and its sex. The disciplining of gender is dependent upon visual and acoustic surveillance systems responsive to ideas about hygiene. Bodies are separated and subject to quarantine under the auspices of health and safety. 'The superego of the hygienic movement constructs modernity by plumbing the destructive instinct of the pleasure we take in dirt and pollution' (Lahiji and Friedman 1997, 53). The hygienic superego must also involve the regulation of sex in public space and denigrate the pleasures taken in touch and smell. The lavatory is a place where we lose parts of ourselves and deaden tactile and olfactory sensations. But it is also the locale in which we feel the prohibitions on love, desire, and gender identification acutely and are thus compelled to transgress heteronormative and cissexist body politics. If gender is about loss (Butler 2004) and the way we go on desiring in the face of loss, it stands to reason that the bathroom conceals as it also reminds us of what has been lost in the making of the gendered bodily ego. 'All identification begins in an experience of traumatic loss and in the subject's tentative attempts to manage this loss ... identification is fundamentally a *reactive* mechanism that strives to preserve a lost object relation while simultaneously searching for a substitute gratification' (Fuss 1995, 38).

The ethical task is to consider how we may deal with lost objects and culturally inscribed prohibitions in such a way that we do not project our grievances onto those who are at odds with the regimes of gender mandated by transphobic and homophobic body politics. We must plumb the toilet so that it does not function as an architecture of social exclusion. A more fluid plumb line, one that can accommodate a range of gender identities and erotic personalities, is the ultimate goal. A truth about the body and its gender is consolidated by internal (psychic) and external (architecturally imposed) body maps and mirror-like plumb lines. The plumb line is a point of demarcation between the me and the not me in space. The plumb line is an 'instrument to measure and to create a level ground ... [and] a sounding device used by anglers' (Costa 1997, 93). Water is integral to the establishment of a plumb line.

> The plumb line is suspended hovering over a virtual water surface, which in turn seems to await the plunge of a plummet. The basic geometrical concepts of horizontal plane and vertical line are thus incarnate in the material qualities of weight and suspension, and of water stillness and balance. A still water plane reflects images of what is suspended over its surface. The problem of reflection and image is exposed in the mythological figure of Narcissus, whose story and iconography stand for the reflexive construction of identity. (Ibid., 93–4)

The plumb line is mirrorical in its dependence upon the water's stillness. The refraction is illusory because it rebounds off a flat surface, one that does not do justice to the weight, dimensions, or fleshy matter destined to fall (or to break the mirror's stillness). One is captivated by the mirror image (or plumb line). To look down is to feel as though one could sink, or perhaps to feel paralysis or vertigo. The mirroring effect enabled by the plumb line, a gaze into a water-filled sink or toilet bowl, is a projective and fanciful tool. It functions like a screen transferring self-image onto architectural objects and spaces. One *appears* in the toilet, and optics work in harmony with the plumb line. The coherence of the subject consolidated by the plumb line is always in danger of falling, of sinking into or ruffling the waters.

The plumb line orders life, and in doing so depends upon the return of what can be cast asunder. 'The architecture of modern plumbing signifies the apparatus of uprightness, the ethics of the Good, and ordered regulatory systems, which issue primarily from representations of the identity of the Male body' (Lahiji and Friedman 1997, 41). The plumb line extols the virtues of the upright position (logos), and denigrates the horizontal (thanatos), diagonal (erotic), or oblique (feminine) position (Lahiji and Friedman 1997). Modern sewer systems are about the triumph of life over death, order over chaos, light over dark, form over matter, masculine over feminine, and so on. In *Civilization and Its Discontents*, Sigmund Freud argues that human uprightness is about the devaluation of anal eroticism and the privileging of the visual, as opposed to the olfactory perceptual system. Uprightness is also about shame and objectification. Genitals, Freud tells us (2002 [1930]), are made visible by uprightness in ways they were not when we moved on all fours. Panopticism works in harmony with the upright position; when we are erect and standing tall we can be subject to the gaze of others. Our images are intercepted by glass and human mirrors that are (as discussed in chapter 3) triangular and uncanny.

The modern aesthetic preference for vertical lines, drains, sinkholes, and flushing technologies reveals our contempt for remainders. 'The water level signifies the horizontal axis of an ideal ground plane. In the relationship between plumb line and water level, something disappears or is, conversely, leftover. Contemporary critical theorists situate this leftover in the notion of *informe* and abjection' (Lahiji and Friedman 1997, 43). It is important to remember that it is not just excrement that is left behind or flushed away as abject human waste. Modern architectural preferences for right angles and boxlike compartments mirror attempts to do away with people who fall out of a normal gender alignment often policed by cissexuals. The plumb line establishes a vanishing point. Those who are perceived to be in excess of the plumb line (masculine women or feminine men, for example) are designated

– literally, by signage on bathroom doors – out of place. So are the queer and erotic uses of the toilet. Those who defy the hygienic superego and the rigidly gendered positions it mandates through lavatory designs, are symbolically coded as abject.

Bathroom architectures are based upon vertical lines and a wish to straighten things out. 'This verticality consists in its obstinate repression of the abject, the unclean … the horizontal' (Lahiji and Friedman 1997, 8) along with visibly queer and/or trans people. Body remains (tears, ejaculate, shit, piss, spit, urine) are flushed down the drain while LGBTI folk are excommunicated (removed by force) or subjected to gender panic. To plumb is to make straight or to sound out, while to plummet is to fall (or to take the plunge). To sink in the transitive construction is

> to cause to descend beneath a surface or to force into the ground; to reduce in quantity or worth; to debase the nature of something, to degrade it, ruin it, defeat it, or plunge it into destruction … [It also refers to] a place regarded as wicked, corrupt, or morally filthy … [The sink] operates in the field of signifiers that indicate a change in state during which the horizontal overcomes the sublimation of the vertical and emancipates the repressed condition of the body from the forces of gravity. (Ibid., 39)

In other words nothing is ever finally (really) abjected (flushed or washed down the sink). There is a metonymic relationship between the abject (that which can be spat or shat into a receptacle) and what has been psychically disavowed in the making of the gendered bodily ego. If the bathroom is a place of forgetting, it may also be a museum or relic of the subject's past. We may say that the toilet stages the primal scene of the abject. With every flush and turn of the faucet we stage a disappearance. Toilet training is about the delineation of the body, its genitals, orifices, and capacities to eject body fluids in time, rhythm, and tempo with a modern capitalist, heteronormative, and cissexist body politic. But it is also about the gendering of the bodily ego. As adults we forget losses incurred by the self, but we are, like the Freudian hysteric, troubled by reminiscences; recollections of the past and renounced desires.

Epistemology of the Plumber

> If I could not be a Prince, I would be a plumber. (Edward, Prince of Wales, 1871)[1]

As Lahiji and Friedman (1997) write in their discussion of plumbing and modern architecture, 'plumbers travel between purity and abjection. They order everyday fluids, manage flow, straighten things out, keep things clean, sound depths, right columns, fix pipes: plumbing leads to the bottom of things' (11). The plumber is employed to unplug (never to dam up). The plumber allows us to drive abject remains underground. Much like the janitor who cleans up the remains of others, the plumber is a custodian of the plumb line. But he or she is something more than this. The plumber is well positioned to be a cultural critic because he or she speaks of the unnamable human soils of the contemporary period. The plumber sees what is left behind. He or she sounds out the depth, or nadir, of the pipes so that abject waste can be funnelled out of the self and down below ground.

But the plumber is not a magician. Despite the most exacting right angles and vertical pipes, the plumb line must give way. There is always an internal limit. 'Our bathroom fixtures might well be our weakest point' (Mallgrave 1997, 19). There is, in other words, a limit to what can be known about the abject. Much like the navel of a dream (which Freud tells us is unplumbable), the toilet (or plumb line) conceals a block. 'In such soundings the plumber eventually reaches an unplumbable limit, a spectral and inaccessible "object" beyond which he or she can reach no more. Plumbing attempts to loosen or unsettle this traumatic kernel' (Lahiji and Friedman 1997, 7). The omphalos, like the toilet, 'signifies that the centeredness of human existence is constructed over a gap, a fissure, a void ... a maternal emblem' (Bronfen 1998, 18), a scar of dependence, rendering the toilet a memorial to what has been forgotten. Trauma is often about the loss of the other. The lost object cannot be returned and is culturally prohibited. 'For the hygienists, shit was the site of irredeemable, even incommensurable loss, which they were obstinately bent on denying' (Laporte 2000, 124). Unlike the hygienist (who instructs others in the art of sanitation – and rarely gets his or her own hands dirty), the cultural plumber navigates the contested terrain of the sacred and the profane, the clean and the unclean. The 'cultural plumber' (Lahiji and Friedman 1997, 11) (like the Freudian interpreter of dreams), be he or she a bathroom patron, scholar, or employee, can tell us something about gender panic and its residue.

Those who are upset by the presence of trans people in toilets do not always know why they are upset. Nor do they understand how LGBTI folk in rigidly gendered and heterosexist space represent an identificatory rather than a physical threat. Gender identifications forged upon repudiated homosexual and cross-gender affiliations are sometimes undone by those who are seen to be queer and/or trans. The return of something cast asunder, or rather abjected, does not come without trouble. What seems to be a refused identification may also be a disavowed identification: an identification already made but forgotten or unknown. In either case, cissexist beliefs in the superiority or authenticity of genders authorized by doctors at birth – their stasis and unchanging moorings, as determined by the body corporeal (its genitals, hormones, and chromosomal counts) – are called into question. Trans folk are often met with suspicion and hostility by cissexuals when their trans status is apprehended. Gender variance is sometimes read as solicitous by those who are not trans because the provocation to question the gendered stasis of the body corporeal is felt to be an inappropriate advance (or invitation). It leads to what Deborah Britzman (1998) refers to as 'difficult knowledge'. It is not that knowledge about gender variance and queer desire is foreign or unfamiliar to the defensive or reactionary cissexual subject ensconced in heteronormativity; indeed, it is all too familiar, and emblematic of unconscious identifications already made (and sometimes even conscious refusals to live and act in nonnormative ways). It is, rather, that the time of the present is not always the time of the past – the unconscious inverts as it also refuses to tell time along a chronological and developmental sequence – and people (not only Freudian dreams) bring us back to early pre-Oedipal scenes, to feelings or wishes, driving identification and desire, that we may prefer to forget in the act of waking (or actively refuse in the wakeful state).

Identifications are unstable and moored by others in space. As Diana Fuss (1995) observes, 'Identification thus operates typically as a compromise formation or a type of crisis management' (49). Difficulties with gender are projected into public space in much the same way as they are fought out in the psyche, which itself operates in multiple time zones. In the Freudian analytic of identification, the unconscious is a 'field of divisions, hostilities, rivalries, clashes and conflicts ... the ego patrols the borders of identity by means of a policing mechanism of its own: identification. Those objects that cannot be kept out are often introjected, and those objects that have been interjected are frequently expelled – all by means of the mechanism of identification' (ibid.). According to this schematization, this is why our identifications are tumultuous and unstable even as they may appear to be coherent and static in adult time. Trans people are appointed to absorb the identificatory crises experienced by cissexuals. Gender policing in the lavatory is not unlike the arbitration and government of conflict in the self. Humane gender recognitions are bound up with how we are being asked to manage our own identifications. What trans folk have to teach or demand by way of humane address, which at the very least involves a respect for one's gender identity, is subject to psychic interference: 'Something inside interferes with the limits of consciousness and the ego's strategies of perception' (Britzman 1998, 7). What is to be learned with respect to gender is difficult to comprehend because it interferes with who we take ourselves to be in relation to others and with our investments in heteronormativity – our toilet training was gender specific, and so many of us were taught to grow up on the straight and narrow.

Normalization is a 'conceptual order that refuses to imagine the very possibility of the other precisely because the production of otherness as an outside is central to its own self-recognition' (Britzman 1998, 82). Likewise, to recognize the other in the self is disorienting – this is particularly true in the domain of gender. The sounding out, or need for clarity about, the gender of others is felt to be primal. Gender identities, like memories, demand that we ratify particular versions of ourselves and others, and this involves a conceptual and psychogenic shift when those identities are at odds with heterosexual matrices. This is true for those who are trans as it is also true for those who are cissexual and invested in heterosexist body politics. Most people want their gender identities validated in public registers. We form attachments to the modes by which we are addressed even when the modalities of address set parameters on what we can be and desire in the social landscape. Because gendered spatial designs policed by cissexuals can be injurious to those whose identifications are at odds with the conventions of gender mandated by those designs, they are the focus of much protest and political organizing as exemplified in the data set.

The call for gender-neutral lavatories is about a wish to validate gender identities that are not always legible in transphobic and heteronormative mirror circuits. As interviewees insist, there must be gender-neutral bathroom options for those who do not fit into the normative gender order. As one non-trans queer femme interviewee says, 'we need unisex bathrooms … so that people don't have to go in and get policed.' As David, who is non-trans and queer, contends, 'I see the most urgent [political] issue being better accommodation for trans people in washrooms … gender-neutral … or even single-user bathrooms are more accessible spaces.' An interviewee from the San Francisco/Bay area says that she 'wouldn't gender the bathrooms …

I think there should be more focus on accessibility and less about surveillance, or secrecy.' Frieda, who is a non-trans queer femme, notes that she is not 'a hundred per cent ... fine [with the idea of degendering all lavatories] ... I know there will be discomfort ... [But] ... it's a human-rights issue ... my discomfort is so insignificant in comparison ... I don't think having gender boundaries does us much good.' Madison, a non-trans genderqueer from Brooklyn, would 'restructure bathrooms so that men and women ... weren't segregated, and [so] that trans and intersex people weren't invisible or left out.'

Many interviewees, however, believe that we cannot do away with gendered bathrooms entirely. Some are reluctant – and often for good reasons – to degender lavatories. Some worry that degendering rooms *en masse* will alienate those accustomed to 'women only space.' One interviewee says that 'gender-open policies in toilets would [not necessarily] create more tolerance. I think initially it may open more people to the strangeness of the "other," which may alienate all concerned and possibly lead to violence.' Perhaps the most frequently cited reason for maintaining gendered bathrooms has to do with the opportunities it provides for gay male public sex. Ivan, who is gay and cissexual, notes that he is a 'little ambivalent about gender-open bathrooms ... because ... [I] think about what I would lose. What some gay men would lose ... the male space for play in a male-specific bathroom ... It [might] not disappear but it would change ... I think the so-called unauthorized uses of bathrooms are important.'

Many interviewees are attached to gender-segregated lavatory designs but agree that the signs on bathroom doors must allow for asymmetries between sexed embodiment and gender identity. If, for example, we take the prototypical stick figures (symbolizing 'man' and 'woman') to be permanently unclear signatures – necessarily at odds with bodies and gender identities (our own and those of others), if it can be permanently unclear what the gender signage means to anyone in particular, and if we can recognize the signage as a mooring or point of departure (as opposed to a definitive locale that can be subject to bio-political regulation) then it may be possible to work creatively with the gap between gender identity and the sex of the body intercepted by others in new and unprecedented ways.

Toilets may offer new and previously unimagined ways of thinking about gender in space. The hope would be to build inclusive spatial designs by understanding what has been denied or foreclosed in cissexist and heteronormative mirror circuits. Spatial designs should not set parameters on who we can be and how we can identify or desire in the social landscape. Rather, spatial designs should cultivate queer and trans-positive fantasy formations. 'Fantasy is what allows us to imagine ourselves and others otherwise. Fantasy is what establishes the possible in excess of the real; it points elsewhere, and when it is embodied it brings the elsewhere home' (Butler 2004, 216–17). Gender may be thought of as a fanciful engagement with what delimits and sets parameters upon who we can be and desire in the social milieu. It is familiar and strange (uncanny), as it is also central to the bodily ego. There is perhaps nothing more personal than the way we seal and delimit the body, and so it is best to leave gender – as an anxious emblem and symptom of subject differentiation – to its own (very often individual) linguistic and psychic devices. If gender, as Butler (1990) suggests, is about how we negotiate love and loss, desire and prohibition, then it must also be structured by trauma. 'To the extent that identification is always also about what cannot be taken inside, what resists incorporation, identification is only possible traumatically. Trauma is another name for identification, the name we might give to the irrecoverable loss of a sense of human relatedness' (Fuss 1995, 39–40).

We become gendered at around the same time as we learn to use the potty. The loss of masculine or feminine ways of being in the world is associated with the toilet because those ways are uncanny and immemorial. Gender, like a navel or omphalos (a conical stone representing the earth's navel) revealing a split or cut at the heart of the adult subject, commemorates loss, or a recess, at the centre of the subject. Gender enables a positioning from which the subject can desire despite having been severed. Unlike the body and its genitals, the navel is not easily gendered; it is an equal-opportunity scar or sealed orifice. The wish for gender-neutral bathroom space is less a wish for a world without gender than a desire to do justice to those who have been socially coded as abject agents or carriers of the losses we all incur in the course of maturation and to whom public space has been denied.

Elisabeth Bronfen (1998) writes that the construction of femininity authorized by the Oedipal story not only translates femininity into an enigma but is symptomatic of how the 'masculine subject projects the recognition of mortality and fallibility' (17) onto the feminine subject. The 'phallic narrative represses this traumatic knowledge by deflecting all the values connected with the paradigm of mortality onto the sexually different feminine body, finding its oblique articulation there' (ibid.). As argued in previous chapters, the phallic narrative tends to position recognizably queer and/or trans people within the domain of the feminine, within the terrain of the castrated and ill-formed subject. Attention to the gendered mechanics of evacuation in the toilet is less about health and safety (as hygienic superegos may imply) and more about the insistence upon the feminine or gender-variant subject standing in as a receptacle. The lavatory is a gendered architecture sometimes imagined to be a womblike or choric maternal space. It is designed to conceal and to serve as a repository for loss (abject remains and disavowed desires). Those losses are projected (sometimes violently) onto feminine, queer, and/or trans folk, albeit in different ways and towards different ends.

But the pedagogy and politics of the gendered toilet are queer and trans-generative. On the one hand they reproduce the binary gender order (and police our positioning in relation to that order by signing doors in ways that cultivate cissexual privilege and gender-segregated space); but they can also undo the foundations of this order. We are never entirely sure who belongs where and why because the gendered toilet is an intermediary space that is strangely familiar or private yet also unfamiliar and public. There is always a gulf between the body and the signs used to signify it in the social field. Accessibility is thus, not surprisingly, imagined to be about the eradication of restrictive gender signs. Nandita, who is non-trans and bisexual, says that she 'wouldn't give [bathroom doors] a label. I'd just ... [call] them bathrooms.' JB, who is genderqueer, notes that there is 'no ... need [to put] a dress or pants on the door.' Haley, a non-trans queer femme, says with exasperation that there should be 'no ladies' signs' on lavatory doors. Phoebe, a queer transsexual woman, believes that we should 'take off the little stick man and stick woman with her skirt, and put *washroom*.' Jacob, who is genderqueer, insists that public washrooms should 'be open to everyone' – and signed accordingly.

Many interviewees wished to subvert conventional gender signage to make room for LGBTI people. Ivan says that the 'symbols [on toilet doors have] ... become so normalized ... The dress for the women, the stick man ... It's infantile when you think about it ... I've seen little spoofs of the male and female symbols reworked, graphically, to depict different things. Like two men butt-fucking ... or two women having sex.'[2] Sasha, who is transgender, says, 'I would label the bathrooms with descriptions that divided people up into nonsense categories. At least that way I could go watch and be amused at people sharing in the same discomfort that I go through [regularly].' Rocky, who is genderqueer, agrees: 'I [would] like something confusing ... a person could go wherever and ... [the bathroom would] not be regulated by a specific norm.'

The Ideal Bathroom

> [I]t can be loudly huzzaed that at the turn of the twenty-first century there is as great a Renaissance in lavatorial design as there was at the turn of the twentieth century, when Britain ruled the sanitary wars. (Lambton 2007, 27)

> [W]hy would the problem of identification not be, in general, the essential problem of the political? (Philippe Lacoue-Babarthe, as quoted in Diana Fuss 1995, 164)

While the vast majority of interviewees questioned the need for gender signs, there were different opinions about how to design the toilet in an ideal world. Trans interviewees often express a desire for enclosed lavatories. For example, JB says, '[I would like to have my] own compartment, so you're not worried [about] ... who's looking under to see if it's occupied, [or think about] who's looking over ... If [the stall] ... was its own solid kind of square room, not square necessarily ... but enclosed [that would be ideal].' JB also says that it would be 'absolutely fantastic for everyone to have their own room ... you could enter a room, and then go into a sub-room, a sub-cubicle ... [the ideal bathroom would] be luxurious ... [It would] have lovely plumbing ... people could enter and exit without it being an attention-drawing situation.' Rohan, who is transmasculine and butch, notes: 'I would love to see a row of twelve doors that are all gender neutral, physically accessible, have change tables, a place to sit down if you want to retreat from the world and nurse a baby.' Nandita says, it's 'pretty cool to be in [an enclosed] bathroom by myself ... no one's going to walk in.' Madison says that her ideal bathroom would be a 'series of specific alone doors ... accessible to everyone.'

Interviewees wanted safety and containment, but also communal and non-restrictive designs. Many people questioned the wisdom of isolation and compartmentalization. For example, Zoe observes that private compartments are 'atomizing us in a way ... [that is] isolating us from each other [in public space]. Maybe that's not a good thing.' JB agrees:

I definitely wouldn't want [this research] to be used to further segregate [gender-variant people] into tighter margins ... As long as there's men's and women's washrooms, then to have something different means that you feel yourself different ... I just think that sometimes in our efforts to treat people 'special,' we end up falling right off the mark by further marginalizing and segregating, and I think that's a bad thing.

Rocky makes a similar comment about privatizing spatial designs: 'I know economically it's not feasible to have single stalls ... for cruising [for sex] ... it's not the best idea, and it's isolationist, which is, like, part of the issue.'

It should be emphasized that the wish for gender-inclusive spatial designs is not incompatible with a desire for a more kinaesthetically pleasing architectural design. As discussed in previous chapters, public facilities are typically built upon vertical and exacting right angles, symmetry, and internally divided and segmented geometrical compartments. Interviewees suggest that the uniformity of public-lavatory designs (white tiles, straight lines, metallic surfaces) produces a cold, clinical, and isolating environment. Such designs symbolically mandate gender purity and a sterile (antisocial) homogeneity. Crystal, who is of Chinese descent, living in Canada, and a non-trans genderqueer woman, says that 'lots of colours, like rainbows, bright colours ... [would definitely] add to the welcoming of the space.' Her vision of the ideal lavatory also involves 'fresh air' blowing in from windows (no air fresheners), accessible 'sit-down and squat toilet' designs, and lots of mirrors. Syd, who is genderqueer, points out that 'toilet spaces are either really narrow ... [or] not really open ... people don't design ... spaces thinking about anyone other than [them]selves.' The ideal bathroom should comfortably accommodate a range of body types, sizes, and abilities. Kew, who is genderqueer and Hong Kong Canadian, elaborates upon an ideal of gender inclusion:

I've actually been imagining ... that [this study would lead to] an installation in a gallery ... a public experience ... I would love to see and participate and be part of something that's experiential ... a travelling exhibition of ... people's stories being told and represented ... through film or audio or ... re-enactment or theatre ... It [would be] all about making bathrooms gender neutral and removing other gendered violence away from bathrooms.

The need for alternative imagery and heterogeneous receptacle designs is often emphasized by interviewees. Alternative designs are associated with both access and luxury.[3]

In her critique of masculinist architectural design and the sexed subject, Luce Irigaray contends that we need heterogeneous designs that do not reduce the sexed subject to an absolute form or shape. Modern engineering and architectural design need to facilitate 'physical and psychic ways of living together with other people in society ... rather than [mimicking] logical representations of the same repetitive properties that fix sexed subjects into reductive formal languages or architectures' (Rawes 2007, 83). Haptic encounters and intimate (relational) spaces are advocated by Irigaray and also by many interviewees. Participants who use the 'women's' room often want 'vanity couches' (places to sit, to lie, and to have sex), circular shapes and diagonal lines (cultivating a warm and gentle invitation into the space), baby change tables, 'family bathrooms' that are not exclusive to those who are part of the hetero-nuclear family ensemble, and, finally, multi-textured and kinaesthetically intimate materials – soft as opposed to fixed (and unchanging) surfaces, gentle and sensual contours as opposed to hard industrial lines.

Interestingly, for interviewees using both rooms, access and luxury are often measured in terms of the familiar or 'homey.' For example, as one trans two-spirited interviewee explains, 'I would probably just have "public bathroom" or "washroom" on the door. It would be open to both genders ... it would ... be very much like a *home* washroom.' One trans male interviewee says that the 'ideal ... public toilet is just one room and you're in there by yourself. It's not a room with multiple stalls, multiple toilets, and a bank of urinals, it's just like a room ... at *home* with a toilet and a sink in it' (emphasis added).[4] It must be noted that familial intimacies and erotic bonds often confuse who we take ourselves to be; we are never sure who we are in the face of a lover or family member. Home is a material and conceptual locale. If public space can be made to feel like home, then it is possible to imagine a gender-inclusive lavatory design that is familiar and yet strange; intimate and yet unfamiliar. Those designated 'other' (outsiders) may come to feel like kin (insiders).

In a similar vein, Irigaray stresses the need for building designs to cultivate relationships between individuals, what she calls 'sexuate cultures' that express and amplify sexual (or sexuate) difference. A shared and intimate environment should 'respect the contribution and place of "others" involved in the production of the built environment' (Rawes 2007, 82). While we may question the extent to which Irigaray's imagining of sexuate cultures is trans-inclusive (and, in fact, whether or not it leaves room to imagine the masculine subject as a sexed, as opposed to an 'unsexed,' subject), I suggest that her analytic of space is useful because it counters a modern tendency to stifle, or in fact to eradicate, the feminine, which, for Irigaray, stands in for sexual (and I would add gender) plurality. If we are to reconfigure the toilet so that it does not institutionalize an oppositional form of sexual difference and exclude people based upon one's access to cissexual privilege, then we must consider how alternative trans-positive designs may prompt a shift in the way we think about gender. Speaking of modern architectural design, Fred Rush (2009) notes that it sometimes

> subverts normalcy in the hope that standard categories of embodied experience will be replaced with new ways to experience the world spatially. It can be profoundly disorienting, requiring the experience of radically non-standard architectural spaces and, in order that experience be critical, a linking up of it with the 'ordinary' by way of memory and imagination, where the possibility to remember the old is always threatened by the novelty in question. (51–2)

By juxtaposing what Rush refers to as the 'known' and the 'unknown' we may produce radically disorienting sensory experiences in bathroom space. But the 'disorientation' – the putting of one's memories and sensory customs 'out of balance' – is not an end in itself. 'The guiding idea is to open up new ways to experience that follow upon an initial disorientation' (ibid., 51). Gender must be rethought along sensory lines and in ways that do not alarm or upset a patron unaccustomed to non-normative architectural designs. We need pedagogically thoughtful lavatory designs that will gently guide unsuspecting patrons through non-normative spatial maps. People must not feel as though they are in danger of being undone or that they may sink downward into an unplumbable or vacuous recess; they should feel as though they are being thoughtfully guided through a passage that is not demarcated by an exacting and excommunicative plumb line. The hope would be that bathroom users may ultimately enjoy another route into a gendered and sensual spatial terrain, one that would be curious and new, physically pleasing, and thought-provoking.

The Urinal in Memorial

Not surprisingly, talk about ideal lavatory designs exposes tensions between communal and compartmental (or, rather, isolationist) designs (as discussed above) and also about the gender of urinals. The standard urinal admits bodies into the province of masculinity, but only if one can use it by adopting an upright and straightforward stance. The receptacle was felt to be either rigidly gendered and exclusive by design or, conversely, a thing of beauty. For example, Cole, who is two-spirited and trans, says, 'I don't think you need a urinal in there, the toilet seat works fine and it's fine at home.' Bryan, who is transmasculine, says with emphasis, 'I think death to the urinal, or at least the unenclosed urinal ... That has to go.' JB concurs: 'I don't understand why we can't all have the same receptacle and have it designed [so] that it works for everyone.' Crystal has a strong preference for unisex toilets but is unsure about the urinal: 'I don't know about urinals ... the urinals are still like ... men taking up this main open space. Maybe the urinals can be in their own room, like each of them in one room, or in ... [a] larger room [where they are less obtrusive].' Kew, on the other hand, refers to the urinal as a thing of beauty: 'I love the idea of urinals. I would keep them, and people could use them if they wanted to.' Several non-trans gay men experience the open-concept urinal as homoerotic and did not want it enclosed by partitions or removed. One interviewee notes that some urinals are almost artful – like water fountains in large city parks – and should be used to grow ferns.

Unlike the oval pedestal enclosed by partitions, the urinal is open concept and may, if differently designed (or staged), invite trans and/ or female and feminine others to imagine themselves in (or before) what has traditionally been a masculine totem or artifact. The urinal is a curiously public yet also a private monument. It is sacred and profane; a thing of white, aesthetic beauty and a receptacle for the grotesque. The urinal also symbolizes the 'upright' position, and is thus dually coded as a white modernist (Freud would say 'civilized') artifact *and* masculinist totem. While looking 'downward' is a heteronormative prohibition (as discussed in the previous chapter), it is also

> equated with losing control, to losing one's stand. Looking downward, one is captivated and paralyzed by the space one occupies. This may be termed as an experience of *distich* space, a space in which the observer is not the active generator of a perspectival order, but rather a stain in a space defined by light. (Costa 1997, 99)

To look downward is to forego a normative, masculine, upright (or straight) stance and to risk proximity to the maternal-feminine (which is horizontal and ill formed) or to the homosexual-sodomitical (which is bent over and racialized as 'dark'). For the masculine subject to remain intact, he must stand upright and appear to be impervious to others. Obsessive fears of mixing are seen before the urinal. The urinal enables one to cleanse the self of abject body fluids. Piss, ejaculate, spit, and vomit are given up. Looking down is about the recall not only of what has literally been pissed or ejaculated away but of that which we lose by way of gender identification. 'Identification is, from the beginning, a question of relation, of self to other, subject to object, inside to outside' (Fuss 1995, 3). Gender is consolidated by posture and positioning in space. A downward gaze at the urinal interrupts a white, heteronormative masculine stance. To lose one's stance, to decentre a straightforward gaze, is to lose one's ground or to reveal that one's identification with masculinity is incomplete or open to resignification. To be beaten in the toilet is often to have one's head shoved into the toilet bowl or to be left in a horizontal position. To be face to face with the urinary or faecal remains of others is to have one's subject integrity undone. What we do before the urinal tells us a lot about gender panic. What (and who) we incorporate and eject is symbolically re-enacted and memorialized at the urinal and in the oval toilet bowl. Excorporation (shitting) is, of course, the antithesis of incorporation (eating), and so it should give us pause to wonder how our gender identities are structured by what (and who) we refuse and let into the terrain of the self.

It is worth noting that female urinals like the Femme pissoire by Chateau Marmont, the Lady P by Gustavberg Sphinx, and the Lady Loo by Goh Ban Huat Berhad have been under development since the late twentieth century. While only the former allows 'women' to assume an upright position (the latter two designs require women to back into urinating positions in more or less squatting positions), they reflect a desire to reconfigure toilet designs so that female and feminine bodies are not required to assume subservient urinary positions. Referring to the Femme pissoire, Barbara Penner (2009) explains that glass mirrors are positioned at eye level around the open-concept female urinal so that women can watch 'themselves standing erect in the act of urination, [and so that] female users may see in the mirror the possibility of reconfiguring these relationships and of reshaping their selves' (148). Similar designs like the L'urinette (urinal for women) have been installed in the Whiskey Café in Montreal, Quebec. It is equipped with directions for use, allowing 'women' to use it 'Just like the boyz!' While they did not create a female urinal, Fierce Pussy (a collective of queer women who do politically motivated public art in the United States) did a graffiti-based bathroom installation at the Lesbian, Gay, Bisexual and Transgender Community Center in New York City that asks users, 'Are you a boy or a girl?' The question is written on the bathroom wall over and above an image of a class photograph featuring elementary school students in Manhattan in 1969. There are, of course, numerous other examples of LGBTI community activism, performances,[5] art installations, posters, and so on, in and about Canadian and American public restrooms that invite us to think critically about the gender of toilet designs.

If a visual economy of disciplinary power is operating to produce a truth about sex – a binary gender division anchored to a so-called truth about the body (as discussed in chapter 3), then mirrors, along with other architectural objects and choreographies deployed by institutional design, must be reassembled. Gender purity is accomplished by optical and, to a lesser extent, acoustic designs, and so it shouldn't be surprising that alternative angles and sound choreographies may invite new ways of thinking about who we are and assume others to be. Recognition is not always more but is sometimes less humane than misrecognition. Recognition (in the Lacanian analytic frame) is dependent upon an externalizing and fantastical *misrecognition* before the mirror – whether those mirrors in adult life are made of optic or auditory nerves, glass, or refractive metal tiles, they are illusory (however important to bodily ego functioning). While there is often pleasure in recognition, it must always be founded upon a case of mistaken identity. There is no truth about the subject to be intercepted, although there *are* identificatory fantasies that are essential to life. The bathroom differently configured could be a place where we build fluid plumb lines, ones that do not impose rigid and exacting visual cuts at odds with internal body portraits. I submit that a luxurious and accessible toilet is a humane project for the twenty-first century.

Glossary

SHEILA L. CAVANAGH AND MELISSA WHITE

> People everywhere categorize themselves and others; this is one of the most fundamental aspects of human language and meaning making. But the ways in which these categorizations are made, and which categories come to have effects in the world, are never neutral.
> (Valentine 2007, 5)

> Because 'transgender' is a word that has come into widespread use only in the past couple of decades, its meanings are still under construction.
> (Stryker 2008, 1)

butch: Refers to those who express masculine traits, mannerisms, and/or appearances. It usually refers to those who are masculine women, transmasculine people, and/or lesbian. Some butches consider their gender identities to be genderqueer or trans. Trans men can also identify as butch, as can non-trans gay men. It is not an exclusively cissexual identification.

cissexual: The prefix 'cis' in Latin means 'same side.' 'Trans' means 'opposite side.' Julia Serrano (2007) popularized the term 'cissexual' to describe 'people who are not transsexual and who have only ever experienced their subconscious and physical sexes as being aligned' (12).

female-to-male (FTM) transsexual: The acronym indicates the direction (from female to male) of the transsexual transition.

femme: Refers to those who express 'traits, mannerisms, or appearances usually associated with femininity, particularly when expressed by lesbian women or gay men' (Stryker 2008, 23). Some femme women consider a femme gender identity as simultaneously genderqueer and/or – not without controversy – trans (i.e., female to femme). Trans women also identify as femme, and it is important to note that it is not an exclusively cissexual gender identity.

genderqueer: Refers to those who do not identify themselves unambiguously as men or women (male or female, masculine or feminine), and/or those who are gender variant. Many genderqueer people have an affinity with queer politics and are critical of binary gender categories.

intersex: Describes a range of physiological variations affecting reproductive anatomy – chromosomal, genital, and/or secondary sex characteristics – making unambiguous gender assignments difficult and/or impossible. Unlike transgender, intersex is often said to be a 'condition' as opposed to an identity. Biological conditions that may lead a doctor to make an intersex diagnosis may include any or all of the following: (1) chromosomal variations on the typical XY (male) or XX (female) karyotypes; (2) genetically male or female bodies that look female or male, respectively, at birth. For example, some female (XX) bodies are born without female reproductive organs (womb, ovaries, vagina); (3) the presence of genitals that have both male and female characteristics (see Stryker 2008, 9). Such conditions are, not infrequently, surgically 'corrected' to enable doctors to make a gender identification at birth. Tragically, this often happens without the knowledge and/or consent of the parents or person upon whom the surgeries are performed.

male-to-female (MTF) transsexual: The acronym indicates the direction (from male to female) of the transsexual transition.

pre-operative transsexual: A person who self-identifies as transsexual and is planning to undergo, considering, or wishes to undergo medical intervention or medically assisted transition.

post-operative transsexual: A transsexual who has undergone medical intervention or medically assisted transition.

transfeminine: Refers to a range of trans people who identify as 'female,' 'feminine,' or 'women,' but who were assigned a 'male' gender identity at birth.

transgender/trans: Refers to 'people who move away from the gender they were assigned at birth, people who cross over (*trans-*) the boundaries constructed by their culture to define and contain that gender' (Stryker 2008, 1). Transgender is an expansive category that can include all gender-variant people.

transgenderist: Refers to those who are transgender but do not wish to have sex-reassignment surgeries. They may, however, take hormones.

transitioned female: Refers to those who have completed a transsexual transition from male to female with or without sex-reassignment surgeries and/or hormones. Those who have completed the transition may refer to themselves exclusively as female or as women (and not necessarily as *trans* women).

transitioned male: Refers to those who have completed a transsexual transition from female to male with or without sex-reassignment surgeries and/or hormones. Those who have completed the transition may refer to themselves exclusively as male or as men (and not necessarily as *trans* men).

transmasculine: Refers to a range of trans people who identify as 'male,' 'masculine,' or 'men,' but who were assigned a 'female' gender identity at birth. A Washington-based group called DCATS (dc area transmasculine society) describes its use of the term as inclusive and celebratory, and identifies its membership as including transmen, female-to-male (FTM) transsexuals, genderqueer, and intersex people, drag kings, and gender-questioning and/or gender-variant bois.[1]

transsexual: Refers to 'people who feel a strong desire to change their sexual morphology in order to live entirely as permanent, full-time members of the gender other than the one they were assigned to at birth' (Stryker 2008, 18). Transsexuals frequently undergo sex-reassignment surgeries (SRS) and/or take hormones. But one may be transsexual without either. Due to the high costs of SRS, hormone therapies, and electrolysis, many street-active, under-housed, and underemployed trans people may not have access to surgical and hormonal treatments but nevertheless consider themselves to be transsexual.

two-spirited: A First Nations/indigenous North American identity indicating a range or combination of gender identifications, typically involving the celebration of both male and female spirits in one body. Gender and sexuality are often experienced as intertwined with one's spirit, language, culture, land, and history. Some two-spirited people use the term to demarcate a specifically indigenous identity – one that departs from the categories gay, lesbian, bisexual, and/or transgender that have currency in white (non-indigenous) communities in North America.[2]

Notes

Introduction

1 It should be noted that *The Metamorphosis of Ajax: A Cloacinean Satire* (the first book on toilets written in Britain), by Sir John Harington, who in 1596 invented the first water closet, called the 'Ajax,' was censored because it was deemed depraved and obscene. While Harington's original design had the basic flushing technologies needed to plumb abject fluids into city sewers, it was not taken seriously until at least 200 years later, when sanitary engineers, plumbers, and inventors of the Victorian era saw fit to recast their eyes upon the Ajax.

2 Gender-rights activist groups across Canada and the United States have been responding to cases in which gender-variant people are denied access to public washrooms. For example, the Transgender Law Center in San Francisco released a now well-known document entitled *Peeing in Peace: a Resource Guide for Transgender Activists and Allies*. In 2001 the Law Center also conducted a Gender Neutral Bathroom Survey in the San Francisco/Bay area which concluded that transgender people frequently experience harassment, assault, and discrimination when attempting to use public facilities. The Sylvia Rivera Law Project in New York provides legal services to low-income transgender and gender-variant communities and is also working to educate people about the importance of gender-accessible toilets. See the film *Toilet Training: Law and Order in the Bathroom* produced by the project, under the direction of Tara Mateik.

  The National Student GenderBlind Campaign: Educating, Advocating, and Organizing for Gender Inclusive Campus Policies, prepared a document that includes information about gender-identity-inclusive policies pertinent to washrooms at colleges throughout the United States. People in Search of Safe and Accessible Restrooms (PISSAR), a coalition of students, faculty, and staff at the University of California at Santa Barbara, is perhaps the most well-known activist group working towards the building of physically accessible and gender-neutral toilets in the United States. Donning yellow T-shirts labelled 'free 2 pee' and latex gloves, equipped with measuring tape and questionnaires, and unapologetically public in their assessment of university washrooms, PISSAR activists endeavour to promote universally accessible toilet designs and policies (Chess et al. 2004). College and university student groups across North America are also lobbying for gender-neutral washrooms for students. See, for example, Dorenson, 'Some restrooms at UA will be "gender neutral,' *Arizona Daily Star*, 16 November 2006; 'UGA adds "gender-neutral" bathrooms,' *Athens Banner-Herald* (Atlanta), 18 November 2006; Vacha, 'U of A considers gender-neutral washrooms,' *Gauntlet News* (Edmonton), 23 November 2006; Clark, 'Gay marriage debate,' *BG News*, 22 September 2006; 'Equality in the toilet,' *Eye Weekly*, 4 December 2003; Hui, 'Washrooms for everyone,' *XTRA West*, 24 June 2004; Gedan, 'Group wants transgender bathrooms for UMASS,' *Boston Globe*, 20 October 2002; Wu, 'In trial run, lines redrawn: Gender-neutral restrooms debut in dorms,' *Student Life News* (Southern California), 22 September 2006; Arduengo, 'Students hope to change gender policy,' *Washington Square News*, 5 October 2006; Perry, 'Ivy League schools enact trans benefits,' *Washington Blade*, 13 October 2006; Curran, 'Unisex toilets controversy swirls at McGill,' *The Gazette* (Montreal), 15 November 2003; Alphonso, 'Universities heed the call for genderless washrooms,' *Globe and Mail*, 10 February 2004. At York University, Toronto, Canada, the SexGen committee lobbied the administration to instal gender-neutral toilets. York is among the first Canadian universities to have gender-neutral lavatories with male and female signs interlinked (and prominently displayed) on bathroom doors. Colleges and universities across Canada are now institutionalizing gender-accessible toilets.

3 The two cases receiving the most press coverage in the United States involve Dean Spade and Helena Stone, both of whom were denied access to bathrooms at Grand Central Station in New York. Dean Spade, who works as a legal advocate for transgender people and is an assistant professor at Seattle University School of Law, was arrested for using the 'men's' room. He explained that he was in the correct bathroom but 'understood the officer's confusion, and offered to use the premises quickly and leave' (Nguyen, 'Interview: Forthcoming from Maximum Rock'n'Roll,' 23 November 2004). The officer called for back-up, restrained Spade, and proceeded with the arrest. Helena Stone filed a lawsuit against the

Metropolitan Transportation Authority (MTA) after being arrested for using the 'women's' washroom at Grand Central Station and settled her case out of court (Donohue, 'Girl's room his, too. Transgender men free to use all of MTA's loos,' *New York Daily News*, 24 October 2006). In a precedent-setting agreement, the MTA agreed to allow people to use washrooms consistent with their gender identities, paid Stone's legal fees, and dropped criminal charges launched against her (which included disorderly conduct). Stone, who is seventy years of age, reports that she was verbally harassed by MTA officers, forced to urinate in a cup (because she was denied access to washrooms), and wrongfully arrested by police on three occasions.

4 In Santa Barbara, Riki Dennis, a student in transition from male to female was using the 'women's' room at a highway rest stop when she was attacked by a jealous boyfriend (who assumed that she would proposition his girlfriend in the washroom) (Brown, 'A quest for a restroom that's neither men's room nor women's room,' *New York Times*, 4 March 2005). Kyle 'Tawnie' Riekena of Butte, United States, was transitioning (male to female) when she was physically assaulted for using the 'women's' restroom in a neighbourhood bar. (The Party Palace owner, Ted Deshner, made a public announcement in his bar that evening saying that 'if you have a penis, you should not go into the women's bathroom ... I don't care if you have a doctor's excuse') (Kelling, 'Transsexual gets into scrape in Butte bar,' *Montana Standard*, 10 October 2005).

5 School administrators, parents, students, and community members throughout the United States and Canada are wondering which bathroom transgender teachers should use and what impact this will have upon student gender-identity development (Parry, 'Hate is not a family value: Schools adjust to transgender teachers,' *Associated Press Newswire*, 25 November 2006). In Eagleswood Township, near Atlantic City, parents were concerned about a transgender (male to female) substitute teacher, Miss McBeth. They attended *en masse* a 'school board meeting last winter, some decrying what they termed an experiment, with their young children as guinea pigs ...' ('Transgender Teaching,' *Associated Press*, 23 November 2006). Carla Cruzan, a teacher at Southwest High School (who is not trans), 'filed a complaint with the Minnesota Department of Human Rights ... alleging that her rights to privacy are being violated because a transgender (male to female) colleague is permitted to use the same restroom' (O'Connor, 'Law firm takes on case against transgendered librarian,' *Star Tribune* (Minneapolis), 24 August 1999). Cruzan refused to recognize her colleague as a woman and believed that her privacy and safety were

compromised by the trans-positive school board policy. The *Ottawa Citizen* reported that a transgendered high school student in Nanaimo, British Columbia, received permission from her principal to use the 'girl's' bathroom after beginning her gender reassignment from male to female (Middleton, 'Transsexual teen's use of girls' toilets raises fears,' *Ottawa Citizen*, 5 March 2002, A3). The decision was based on a British Columbia Human Rights tribunal decision in 1999 ruling in favour of a transsexual woman's right to use the 'women's' bathroom in a Victoria bar (Rud, 'Student's Gender Switch,' *Times-Colonist*, 5 March 2002, A1). Parent Advisory Council chair Vicki Podetz, of Nanaimo-Ladysmith School Board in British Columbia, requested that the grade twelve student use a 'gender-neutral' washroom (the toilet designated for students with disabilities) instead of the 'girl's' washroom because the decision did not take into consideration the 'comfort level of the [non-transgender] female students' (Middleton 2002, A3). Podetz insisted that allowing a transgender student to use the 'girl's' washroom compromised the 'privacy of other female students at the school' (Rud 2002, A1). The parent led a small number of Nanaimo-area students, along with four parents, to a protest in front of Cedar Secondary School.

There is now a gender-neutral washroom for students at Park Day School in Oakland, San Francisco. 'Park Day's gender-neutral metamorphosis happened over the past few years, as applications trickled in for kindergartners who didn't fit on either side of the gender line. One girl enrolled as a boy, and there were other children who didn't dress or act in gender-typical ways' (Lelchuk, 'When is it okay for boys to be girls, and girls to be boys?' *San Francisco Chronicle*, 27 August 2006). The California Student Safety and Violence Prevention Act of 2000 prohibits discrimination on the basis of gender identity. This legislation is being used to make bathroom and change-room provisions for gender non-normative children.

A New York organization, Advocates for Children, interviewed seventy-five lesbian, gay, bisexual, and transgender students and learned that many students had to fight to use a school bathroom consistent with their gender identity (Yan, 'Gay students face hostile environment, report says,' *New York Newsday*, 28 October 2005). Numerous other cases in the United States and Canada of student difficulty accessing appropriate washrooms without harassment by students and teachers appear in the news media (Meadows, 'From girl to boy,' *People*, 30 October 2006).

6 Trans workers across North America report that they are unable to access toilets at their place of work (Bentley, 'Transgender worker sues,' *Star Tribune*, 21 January 1999; Brown, 'Transsexual rest room riles T Blue Line staff,' *Boston Herald*, 7 June 2000).

7 Apart from employment obstacles, gender-variant people face infantilizing and unsafe conditions resulting from the policing of gender in public washrooms in all public spaces. For example, Natasha Lee West, a transgender woman, was barred from using the 'women's' change room and told that she should use the 'children's' bathroom at her fitness facility (Jacobs, 'Health club harasses transwoman,' *Bay Windows*, 16 August 2006). West was denied access to the 'women's' change-room and washroom at Bally Total Fitness club in Worcester, Massachusetts, because she had not had sex-reassignment surgery (their prerequisite for a female gender identity). 'West told her [the staff member] that she was willing to forgo using the changing room, but she needed access to a restroom. She said the staff member led her to the gym's daycare center and told her she could use the children's bathroom. West said the bathroom consisted of a small plastic toilet with the seat raised only about a foot off the ground' (ibid.). Later, the fitness staff told West that she would have to use the 'men's' bathroom located in the 'men's' change-room, which, as West pointed out, would put her safety at risk.

8 'Tranny-queer' is used here to indicate how transphobic and homophobic slurs are sometimes combined. 'Tranny-queer' is often used as a pejorative term to denigrate trans women who are prostitutes and/or performers and/or street involved. It has more recently been reclaimed as a self-identification by some trans activists and trans youth. It does, however, remain controversial within trans communities, given its painful and derogatory genealogy.

9 This study focuses on Canada and the United States exclusively. While Mexico is a federal constitutional republic in North America, there are significant variations in toilet design and usage in that country influenced by Mexico's colonization by Spain. Unfortunately, an analysis of Mexican lavatories is beyond the scope of this study. I do, however, hope that others will consider how the gendered politics of access operates in Mexico as well as in other Latin American countries.

10 Washrooms are, as many people say, disgusting abject places. In his discussion of faecal imagery and rhetorics of abuse, David Inglis (2002) notes that socially disenfranchised groups are spoken about in terms of excremental filth. He refers to racial slurs like 'Paki shit' and 'Jewish turd' to illustrate the extent to which the rhetoric of faecal matter is used to align racialized and disenfranchised people with toilet culture. The extent to which social groups are associated with the lavatory tells us something about the degree to which they are vilified. 'Here the subordinate can be depicted as more faecally uncontrolled and excrementally libidinous than

their apparent superiors' (ibid., 208). Marginalized groups are all, in various ways, constituted as abject by linguistic techniques – metaphor, allegory, stereotypical imagery, 'metaphysical condensation' (Morrison 1992), 'metonymic displacement' (ibid.), and 'fetishization' (ibid.). It is often the case that the marginalized group is paired with faecal matter and toilet iconography. The metaphorics and imagery of shit figure prominently in the discursive production of others and otherness in the western-European and North American political landscapes. Inglis (2002) confirms that 'From early modernity, the bourgeois faecal imaginary relied upon and sought out the bodies of groups that could perform the symbolic operations of filthy Otherness' (214).

11 See, for example, the article published in the *New York Blade* (an LGBT community newspaper) entitled 'Bathroom break: Settlement reached in West Village eatery lawsuit; masculine woman was ejected from women's bathroom during Pride 2007' (9 May 2008: 3); and 'Washroom conflict escalates' (1 May 2003) published in *Xtra West* (Vancouver's GLBT community newspaper), about a trans man ejected from the 'men's' toilet in Avanti's gay pub on Commercial Drive.

12 For a discussion of intersexuality, see: Cheryl Chase, 'Hermaphrodites with Attitude' (2006); Anne Fausto-Sterling, *Sexing the Body* (2002); and Suzanne J. Kessler, *Lessons from the Intersexed* (2002).

13 For a critical discussion of the institutionalization of the term 'transgender,' see David Valentine, *Imagining Transgender* (2007).

14 For a discussion of butch identifications, see Gayle Rubin, 'Of Catamites and Kings' (2006).

15 For discussions of female masculinity, see Judith Halberstam, *Female Masculinity* (1998); and Jean Bobby Nobel *Masculinities without Men?* (2004).

16 For a discussion of FTM identities in Canada, see Jean Bobby Nobel, *Sons of the Movement* (2006).

17 For distinctions made between transgender and transsexuality, see Viviane K. Namaste, *Invisible Lives* (2000).

18 Queer theorists have been faulted for not attending to the specificities of transsexual life experiences and transphobia (Boyd 2006; Devor and Matte 2006; Namaste 2000; Prosser 1998a).

19 For a discussion of how public toilets are inaccessible to persons with disabilities, see Rob Kitchin and Robin Law, 'The Socio-spatial Construction of (In)accessible Public Toilets' (2001).

20 In order to abide by ethical protocols mandating that participant identities remain confidential, I have not indicated when a real name as opposed to a pseudonym is being used.

1. Queering Bathrooms: Gender, Sexuality, and Excretion

1 See Peter Stallybrass and Allon White (2007) for an interesting discussion of the London city sewer and how it was used to orchestrate divisions between social classes in the Victorian era.
2 Sewers in Paris were regarded as part of the criminal underworld: 'After mugging pedestrians or stealing from shopkeepers, thieves would slip into the depths of the sewer confident that no one would follow them' (Horan 1996, 91). A mob called the Swamp Angels similarly used the sewers to hide in after having committed crimes in New York City.
3 The word 'orifice' originally meant the 'opening of a wound' in sixteenth-century France.
4 Everyone shits, and it lands in the same sewage tank. The institution of the public lavatory was a place 'where the modern subject must confront both their bodily porousness and their conflicted relation to being, *en masse*, in public' (Morgan 2002, 193). Sewer systems were interconnected underground. The very purpose of the modern sewer system was to funnel human excrement out of homes and into a mass septic system. Urine and faeces are mixed and driven underground via communal pipelines. Symbolic encroachments were felt at the level of the body. Prostitutes sat where ladies rested, and the sodomite stood where a gentleman relieved himself, and so on. Public lavatories brought people of diverse social and class positions together, and so they became architectures in which boundaries and encroachments were vigorously policed.
5 Speaking of two children (a boy and a girl) viewing the signs 'Ladies' and 'Gentlemen' at a station (through a train window), Lacan (2006) writes: 'To these children, Gentlemen and Ladies will henceforth be two homelands toward which each of their souls will take flight on divergent wings, and regarding which it will be all the more impossible for them to reach an agreement since, being in fact the same homeland, neither can give ground regarding the one's unsurpassed excellence without detracting from the other's glory' (417).
6 It is interesting to note that Jeremy Bentham designed modern prisons with visual economies of power and surveillance systems particular to the modern period (as theorized by Michel Foucault in his now famous *Discipline and Punish*), and *also* stressed the importance of efficiency and utility in dealing with physiological functions, thereby influencing hygienists concerned with the disposal of human faeces in European cities (Laporte 2000, 123).
7 Critical geography has, in recent years, been concerned with the colonially inflected spatial production of whiteness (Dwyer and Jones 2000; Nast

1998, 2000), and postcolonial theorists have noted the centrality of the English water closet to processes of racialization by British, European, and American colonial and imperial powers (W. Anderson 1995, 2002; Anspaugh 1995; Chun 2002; Cummings 2000; Inglis 2002; Largey and Watson 1972; Morgan 2002; Srinivas 2002; Van Der Geest 2002).

8 The development of capitalism, industry, technology, and other regimes for monitoring individual conduct correspond with the advent of the English water closet. For example, Dominique Laporte, who wrote *Histoire de la merde* in 1978 (translated by Rodolphe el-Khoury 2000), notes that the toilet corresponds to the rise of capitalism. Inglis and Holmes (2000) also argue that the development of the English water closet corresponds to the more intricate divisions of labour in post-feudal Europe, to changes in the regulation and governance of workers – routine, carefully monitored activities in accordance with a larger, assembly-line system of capitalist production – and, finally, to the creation of a docile, timed, and managed modern body. Capitalist configurations of time, geared towards production schedules, efficiency, and the disciplinary control of workers on factory floors, all had significant impact upon the body, its rhythms, and its excremental patterns. It is worth noting that Sigmund Freud (1960 [1917]) believed that anal eroticism must be sublimated to develop an appreciation for the value of money, commerce, and trade.

9 It must be stressed that toiletry technologies were first built to consolidate class divisions not to guard against disease. It was not until the late nineteenth century that personal hygiene and cleaning rituals, including the flushing away and containment of human faeces, were considered to be important to disease control and human health.

10 For a discussion of how racialized immigrants to the United States who worked in the area of recycling and waste management were regarded as 'dirty,' see Carl Zimring, 'Dirty Work' (2004).

11 Modern European ideas about hygiene and cleanliness date back to the mid-eighteenth century when the aristocracy practised the new art of bathing. 'Bathing conferred status on the bather: It was associated with luxury, wealth, refinement, and self-awareness and was thought to be an attribute of the upper classes' (Srinivas 2002, 379). It was not until the late eighteenth century that bathing rituals and hygiene practices filtered down to all classes of people, who previously had not been able to afford in-house porcelain or metal tubs. In the United States, personal hygiene was a target of social reform (Sivulka 1999; M.T. Williams 1991). As Richard Anderson explains, 'The public bath house or lavatory allowed the reformers to ameliorate the social stigma of dirt by enabling the poor to

keep themselves clean. Accordingly reformers built a substantial number of municipal and charitable bath houses, steam laundries, swimming pools, and public bathing stations, which were usually located in the most congested immigrant neighbourhoods ... stigmatizing the poor as "unwashed" also served to maintain the sense of social distance, and sanitized racial or class prejudices with an aura of medical responsibility' (2008, 3–4).

12 For an interesting discussion of the use of scatological imagery and religious iconography in art, see Susan M. Canning, 'The Ordure of Anarchy' (1993).

13 For a discussion of the symbolic equation of money and faeces in psychoanalytic thought, see Alexander Kira, 'The Bathroom' (1976).

14 In 'Fragments of an Analysis of a Case of Hysteria' ('Dora') (1975 [1905]), Freud writes in a footnote that 'It is scarcely possible to exaggerate the pathogenic significance of the comprehensive tie uniting the sexual and the excremental, a tie which is at the basis of a very large number of hysterical phobias' (63, note 1). While we may question Freud's understanding of hysteria and female psycho-sexual development, it is significant that the cultural association between sex and faeces gains recognition in the early twentieth century.

15 Julie L. Horan (1996) notes that, in seventeenth-century Germany, sexual problems between lovers were dealt with on the wedding night by pissing through the wedding ring: 'Obviously if a male could hit the small space in a wedding ring, he had talent. If a woman wants to end without her having to endure a direct confrontation, she might surreptitiously place a tiny sample of her dung in her unwanted lover's shoe. The smell of her, now subconsciously associated with this bad odor, was supposed to drive him away' (61). In Italy, doctors were employed to determine a lady's virginity by sampling her urine (Horan 1996, 61). In the Middle Ages women made a potion called a 'love philter-aphilter' to make a man fall in love. As Horan explains, 'The potion was made of the woman's own excrement. A remedy to the love potion was concocted of human skull, coral, verbing [sic] flowers, afterbirth, and urine. As a testament to the believed powers of the potion, the use of love philter-aphilter was punishable by death' (130–1). In *The History of Shit*, Dominique Laporte (2000) laments that a history of the cosmetic uses of faeces by 'ladies' is yet to be written: 'Who will write the history of Saint Jerome, advisor to the Ladies of Rome from 382 to 385, who warned against the practice of smearing one's face with shit to preserve a youthful complexion?' Speaking of Sicily's women, Laporte also wonders how we could condemn the practice of bathing and gargling in urine 'when [Italy's] own civilization ... owed the beauty of its women to the shit in which they bathed?' (ibid., 102–3). 'Ladies' kept shit

on their 'dressing-room tables beside their most cherished fineries' (ibid., 106–7).
16 Margaret Morgan contends that 'Modern plumbing, in its connection to woman, has acquired these associations, both revered and reviled. Plumbing is the uncanny embodiment of the sexualized and maternal figure of woman – erotic, comforting, horrific. Tap water is our pure mother, waste water our slut' (2002, 174).
17 The working-class male body was also an object of public concern. This was so much the case that in Dunedin, New Zealand, people lobbied for the building of toilets exclusively for men in the mid- to late nineteenth century: 'After 1860, the influx of heavy-drinking gold miners generated a campaign marked by concern over filth and the management of unruly men, resulting in the first civic provision of toilets for men' (Cooper, Law, and Malthus 2000, 430). Richard Anderson (2008) similarly notes that in Victorian and Edwardian cities (like Toronto) officials tried to regulate the 'urinary behavior of working class men ... through the construction of public lavatories' (3). 'Despite prohibitory bylaws it was an open secret that males fouled the gutter. In 1908, noted one Toronto official with disgust, men were defiling the lawns of City Hall' (ibid., 7).
18 See Gryphius's publication *In latrines mortui et occisi* for a discussion of 'eminent men and women who were born and died in infamous places – namely, in latrines' (Laporte 2000, 36–7). Wallace Reyburn (1969) tells the story of guests drowning in cesspits: 'In 1844 no fewer than fifty-three overflowing cesspits were found under Windsor Castle. There was the case of one titled host, at the front door of his country mansion to greet the arrival of a coach full of weekend guests, being called upon to watch in horror as the driveway subsided and they were engulfed in an overflowing cesspool. And, sad to relate, it was not without loss of life' (25).
19 For a good discussion of the history of cleanliness in France, see Georges Vigarello, *Concepts of Cleanliness* (1988).
20 Sir Edwin Chadwick of London (author of a sanitation report on the labouring classes) advocated for the segregation of the poorer classes (the vagrant and under-housed, the prostitute and degenerate, along with the so-called criminal classes) from respectable white, bourgeois society, by sanitary design.
21 For a discussion of the International Olympic Committee sex tests, see Sheila Cavanagh and Heather Sykes, 'Transsexual Bodies at the Olympics' (2006).
22 For a productive critique of Prosser's analytic of the bodily ego, see Patricia Elliot, 'A Psychoanalytic Reading of Transsexual Embodiment' (2001).

23 For a discussion of how people are culturally coded as abject, see Butler (1990, 1993), Thomas (2008), and McClintock (1995).

2. Trans Subjects and Gender Misreadings in the Toilet

1 See Browne 2004; Davies 2007; Halberstam 1997a, 1997b; Juang 2006; Magni and Reddy 2007; Pilling 2006.
2 It is interesting to note that people with physical disabilities are given their own gender-neutral toilets.
3 In *Undoing Gender* (2004), Judith Butler draws upon Hegel's analytic of desire and recognition to illustrate the extent to which gender recognitions enable each of us to enter into the social field. The categories 'male' and 'female,' 'man' and 'woman' allow some bodies to be recognized and constituted as social beings while preventing others who depart from, do not fit into, or reconfigure gender norms from entry into personage. 'And sometimes the very terms that confer "humaness" on some individuals are those that deprive certain other individuals of the possibility of achieving that status, producing a differential between the human and the less than human' (Butler 2004, 2). When gender categories act as identification papers, credentials, and 'passports' through which we are deemed eligible to enter into a given spatial field – like the public toilet – we are confronted with the difficult questions, For whom is the requisite met and for whom is it foreclosed? Who is able to be recognized and located in a normative spatial field and who is denied access and designated 'out of place'?
4 I am in agreement with Lacanian theorists who contend that all recognitions are, in fact, 'misrecognitions.' There is no essence or truth about gender to recognize. Recognitions are based on illusory and externalizing visual projections of ourselves and of others, made possible by glass mirrors and by those we use as ego-friendly mirrors to consolidate our own internal self-portraits. Mirrors cultivate illusions about the stability of the body, which can otherwise feel as though it is in 'bits and pieces.' It is important to distinguish between a 'misrecognition' essential for one to come into existence as a subject (in the Lacanian sense) and a public (or wilful) 'gender misreading' that blocks one's access to gendered space and public participation. This book is concerned with gender misreading that are excommunicative and transphobic. So long as public space is gendered, it will be necessary to validate gender identifications in accord with individual self-identifications. There is a political need to confer one's gender identity (trans and otherwise), even as these identifications are fictitious and fantastic. It is important to note that those who are transgender are no

more ensconced in culturally salient gender identities than those who identify (and invest in) their gender assignments at birth.
5 This interviewee identifies herself as a 'transgenderist,' which means that she is taking hormones but not wanting sex-reassignment surgeries.
6 See Sheila Cavanagh, 'Teacher Transsexuality' (2003).
7 While the discourse of 'passing' is important to many trans interviewees, Julia Serano (2007) argues that it is limited to the extent that it is an active verb denoting something one does, as opposed to a noun denoting something one is. To talk of the ability to 'pass' is to suggest that trans people are deliberatively deceptive about their gender identities, whereas those who are not trans continue to have an 'authentic' gender status. Secondly, the discourse requires trans people to negate their histories and trans status. 'Cissexuals (not transsexuals) are the ones who create, foster, and enforce "passing" by their tendency to treat transsexuals in dramatically different ways based solely on the superficial criteria of our appearance' (178–9). What many people now refer to as the ability to 'pass' can more accurately be described as the ability to access 'conditional cissexual privilege' (ibid., 180). For another critical discussion of 'passing,' see Mattilda, aka Matt Bernstein Sycamore, ed., *Nobody Passes* (2006).
8 In her discussion of Merleau-Ponty and the specular body, Vasseleu (1998) writes that the postural schema along with the 'perception of others and the sharing of perceptual experience is possible, not because the bodies of others are recognizable as projections of one's own body, but because they have a perceivable comportment and intentionality as relationships that are translatable between bodies' (51). One knows that one can be seen as one sees and that if the other can be gender incoherent and/or trans so too can oneself. The differences between the self (as subject) and the other (as object) are insecure in the visual field – while vision is also, as argued by phenomenologists, the sensory system most responsive to distance and objectification.

If it can be said that those who are cissexual call upon those who are trans to identify with the transitive nature of all gender-identificatory circuits, then it must be agreed that the rigid gendering of the toilet is about the maintenance of psychic and postural integrity for those unwilling to see themselves as shaped by the other. 'What is being considered here is the body folded back on itself in its otherness as a seeable, feelable, audible thing' (Vasseleu 1998, 53).
9 Dorothy, who is white and genderqueer, is one of a few interviewees who forget, or rather get confused about, gender signs on toilet doors: 'I frequently have this moment ... of confusion as I am approaching [bathroom]

... doors, and I am like "Oh God, which one do I go in?" and I sort of have to make myself remember that I use the women's. And I don't really know what that feeling of being restricted is about ... [but there is] something ... visually upsetting about walking up to two doors and having to make a choice and feeling that confusion about how I ... [should identify] ... it forces me to remember that I am perceived in one way or another ... and I don't feel like I fit in [to] one or the other. So it's sort of forcing this identification on me ... you have to make a choice [about gender] ... and it's a choice I don't really want to make.'

Diane, who is a non-trans woman, offers a similar narrative account about reading gender signs on doors: 'I never wear a dress or a skirt and I actually have walked into the men's washroom, which is so funny because I look at that [sign on the door] and it looks like me and I walk right in and I have to walk back out and I can only say that I look at the sign. So if I see a skirt I don't say that's not me and I am not going in there, but it's when I don't see one but I see the other.'

10 Commenting upon another instance, Bryan also relates a troubling encounter with a waiter:

I was travelling ... with my girlfriend and we stopped at ... [a] restaurant diner ... in Wisconsin ... as I went in the door [to the women's bathroom] I saw one of the [male] waiters eyeing me strangely ... the waiter stood outside the [bathroom] door ... loudly debating whether [he] ... should call the police or not and loudly debating who I was and why I was in the restroom and if they should ... come in and haul me out ... and it was really scary [thinking they were going to come in and get me] ... even ... coming out of the bathroom it was scary having to, like, dodge them and get back to the table and get out of there.

11 Nandita, who is a South-Asian non-trans woman, and bisexual, underscores how the gendered and enclosed architectural design of the bathroom is conducive to hate crimes by relating violent incidents: 'Once I went inside [my high school bathroom and] there ... were groups of people waiting for me because they knew that I would turn up there eventually during the day ... [in that] gender-segregated space ... And once you're in there ... it's really [isolated] ... I've tried to make noise but I was prevented from making noise and ... nobody came. And ... other times when I've gone into bathrooms I've found people who have been harassed, with their clothes torn off ... or just really like, insulted, or beaten up. And they're just sitting there, crying. And I've also found people who've slit their wrists in bathrooms ... For me bathrooms are ... space[s] in which people are subject to a lot of violence. And the fact that it's closed

off, the fact that so many people can come in and violate one person and nothing will happen [is upsetting. There is no] ... rescue or support.'

3. Seeing Gender: Panopticism and the Mirrorical Return

1 Ulf Linde first used the phrase 'mirrorical return' in his discussion of the *Fountain*. William Camfield later gave the phrase scholarly attention by using it in his book *Marcel Duchamp* (1989), 106. Duchamp's *Fountain* offered a powerful commentary on the public lavatory and the city drinking fountain, which in the early twentieth century were thought to be infected by germs. Molesworth writes that the 'up-ended urinal look[s] like a fountain, not only is there play with the morphological similarity between a stream of male urine and the jet stream of water from a fountain, and not only are drinking and peeing intimately connected, but both activities had become mechanized, public, and subsequently tinged with dread in the popular print media of the time' (1997, 82).
2 For a discussion of how homophobic hate slips into the terrain of the erotic in *The Toilet*, a play produced in 1964 by LeRoi Jones (Baraka), see José Esteban Muñoz, 'Cruising the Toilet' (2007).
3 This is poignantly illustrated in a story told by an interviewee about a transsexual woman who gave a workshop in Vancouver. While sitting on the toilet seat during a break, the instructor found her body subject to interrogation by a transphobic workshop attendee peering over a stall partition. As described by the interviewee, 'She [the instructor] looked up. And there was a woman from the workshop staring down at her, into the stall. [The workshop attendee] ... standing on the toilet [seat] ... staring down into the [adjoining] stall ... points at her [instructor] and says, "You have a penis." She [the instructor] was, like, "I know ... I just gave you a workshop about it [transsexuality]" [but] ... The woman [student] ... ran out and was screaming.'

The instructor's transsexuality did not evoke a transphobic response until she entered the toilet. In other public venues, transphobic questioning may be more surreptitious, veiled by a pretence of politeness, and less verbose and visually invasive.
4 Most of the transmasculine folk interviewed do not use the urinal in the 'men's' room at all or with regularity. This is likely because most of the trans guys interviewed in my study are pre- or non-operative and therefore have not had a surgical intervention enabling them to stand while urinating. Even those who are pre- or non-operative do not seem to use a funnel or stand-to-pee (STP) device to urinate. Trans men and masculine

people in the *Queering Bathrooms* study often say they need access to a toilet in a closed nook because they prefer or need to sit.
5 Paruresis is a medical diagnosis defined as a 'fear or inability to urinate in public restrooms when other persons are present' (Hammelstein and Soifer 2006, 296). It must be understood within a wider social context, and with attention to the rigid gendering and exclusivity of public toilet designs, as evidenced by interviewee testimonies.
6 Internalized body imagery signifying in excess of the body, its material contours and genitalia (Salamon 2004, 113), is a defining feature of transsexuality. Transsexuality is predicated upon a significant gulf between the image and the sensational or proprioceptive ego. Silverman (1996) notes that the 'proprioceptive ego may not always be compatible with what the reflecting surface shows' (17). To avow this incongruence is part and parcel of what it means to be transsexual. What medical doctors and psychiatrists call 'gender disphoria' may lead one to avow or animate the disjuncture between image and proprioceptive ego: a genderqueer or transgenderist identification (among others) may then ensue. One may also be motivated to reduce a felt lack of synergy between the image and the proprioceptive body because it is uncomfortable or because it produces identificatory dissonance; a transsexual identification may then take shape. But the precarious relationship between the specular imago and the corporeal form (sensation or proprioceptive ego) is something we all must negotiate, regardless of our gender identities (trans or cissexual).
7 The mirror is not a neutral glass into which a 'truth' about the body is apprehended by reflection. It is a normalizing sphere that has already been written and projected upon. The images returned encase the self in ways that are at odds with our own internal self-portraits. By establishing boundaries at the skin, contours and cuts that do not coincide with internal self-imagery, the mirror highlights the fissures between image and sensation, gender and sex, we all avow or disavow in the making of gender identity. The mirror frames and renders static otherwise volatile flesh.

4. Hearing Gender: Acoustic Mirrors – Vocal and Urinary Dis/Symmetries

1 Silverman says that the mother's voice is the first acoustic mirror. Speaking about masculine subjectivity, she suggests that the toddler becomes a subject (and develops his own voice) by hearing the mother's voice and incorporating it as his own, and then, as an adult, projects 'infantile babble' onto the feminine subject. The reversibility of the acoustic mirror facilitates projection and introjection. The masculine subject who first hears

himself through the voice of the mother – he incorporates her acoustic field and sounds his body against hers – is loath to be reminded of his infantile and interdependent origins. The female voice is worrisome because it 'functions as an acoustic mirror in which he [the masculine subject] hears an unwanted part of himself, an element of himself which escapes "social rationality, that logical order upon which a social aggregate is based"' (1988, 82).

2 For a good discussion of modern architecture and critical theory, see Neil Leach, *Rethinking Architecture* (1997).

3 Silverman (1988) defines castration as the loss of an object that 'precedes the recognition of anatomical difference – a castration to which all cultural subjects must submit, since it coincides with separation from the world of objects, and the entry into language' (1). Castration is not about the loss of *a* particular body part such as the penis. It is about a set of losses predicated upon subjective differentiation from a maternal body enabling entry into what Lacan calls the Symbolic through language.

4 See Didier Anzieu's (1989) discussion of the sound envelope for a better understanding of how the 'sound mirror' or the 'audio-phonic skin' plays an important role in the formation of the skin ego. Anzieu notes that the 'auditory sensations produced when sounds are made are associated with the respiratory sensations which give the Self a sense of being a volume which empties and re-fills itself, and prepare that Self for its structuring in relation to the third dimension of space (orientation and distance) and to the temporal dimension' (157).

5 For a discussion of sound and image and gender recognitions, see Susan Stryker, 'My Words to Victor Frankenstein Above' (2006).

6 Lacan defines 'object (a)' (or *objet petit a*) as the 'object-cause of desire.' '*Objet petit a* is any object which sets desire in motion, especially the partial objects which define the drives. The drives seek not to attain the *objet petit a*, but rather circle around it' (Evans 1996, 125).

7 Referring to the 'voiceover' in Hollywood film, Silverman explains that there is a 'general theoretical consensus that the theological status of the disembodied voice-over is the effect of maintaining its source in a place apart from the camera, inaccessible to the gaze of either the cinematic apparatus or the viewing subject – of violating the rule of synchronization so absolutely that the voice is left without an identifiable locus. In other words, the voice-over is privileged [and usually masculine] to the degree that *it transcends the body*. Conversely, it loses power and authority with every corporeal encroachment, from a regional accent or idiosyncratic "grain" to definitive localization in the image. Synchronization marks the

Notes to pages 118–25   243

final moment in any such localization, the point of full and complete "embodiment"' (1988, 49).

8 Recognition in a maternal sonorous envelope is enabled by an acoustic mirror. This recognition comes before the mirror stage theorized by Lacan. Silverman (1988) writes: 'Since the child's economy is organized around incorporation, and since what is incorporated is the auditory field articulated by the maternal voice, the child could be said to hear itself initially through that voice – to first "recognize" itself in the vocal "mirror" supplied by the mother' (80). Hearing precedes vision and occupies an important place in the development of subjectivity.

9 Julie Horan (1996) notes that some nineteenth-century close-stools and night tables played music to camouflage the sounds of elimination: 'One type of close-stool played chamber music when its lid was lifted. A night table provided music when its door was opened for access to the chamber pot. As Victorians found bodily functions embarrassing, the musical cover-up relieved the anxiety of being heard while relieving oneself' (82). There was also an eighteenth-century chamber pot known as the 'jerry' with a 'concealed musical box, that gives a recital of appropriate chamber music when it is lifted, and cannot be turned off by the embarrassed guest' (Wright 1960, 124). It should also be noted that Victorian ladies were uncomfortable with the loud flushing sounds made by the newly engineered water closets. The contraptions made 'flushing,' 'gurgling,' and 'hissing' sounds. The now-famous nineteenth-century sanitary engineer Thomas Crapper even added a silencer to the closet to dampen the alarming sound of the flush. Relatively quiet flush toilets were not manufactured until the late 1900s with the advent of the symphonic closet by Thomas Twyford (Blair 2000). For a discussion of the 'royal flush' and its potential to disrupt the crowning of Queen Victoria, because of the excessive noise generated by multiple flushes in unison at her ceremony, see Reyburn (1969, 22–3). Lewin (1999) notes that in present-day Japan some 'toilet-roll dispensers are equipped with music boxes, which play a tinkling tune, such as the Scottish folksong "Coming Through the Rye"'' (74).

10 Stall doors are also removed by authorities to curtail public sex in bathrooms.

11 Silverman (1988) says that there is a Western cultural association between the vagina (as a 'dark continent') and the maternal voice as sonorous envelope: 'Since the maternal voice is associated with the darkness and formlessness of the infant's earliest experiences, rather than with the form-giving illumination of the *logos*, this anteriority implies primitiveness rather than privilege' (75).

12 The fantasy of the vampiric mother is, for Kristeva, the ultimate example of an abject horror. She is both threatening to subjectivity and also the vehicle through which we come into existence.
13 Referring to a Freudian defence mechanism (called 'sour grapes'), William Miller writes, 'Sour grapes doesn't repress desire and put it behind a barrier of disgust so as ultimately to enhance the pleasure of attainment. It makes the desirable thing disgusting in itself; it kills desire completely; it shrinks the world of possibility, even eliminating the desire to contemplate the grapes – consciously at least' (1997, 116). Desire and disgust are 'two sides of the same coin.' The difference between the two is a question of memory, cultural prohibition, and projective disassociation. Disgust, in Freud, always betokens past and/or repressed pleasure. The same holds for Kristeva's concept of abjection: it, too, connects 'an apotropaic gesture with a prehistoric legacy of pleasure' (Menninghaus 2003, 375).
14 In his colonial account of urinary positions from the Middle Ages to the dawn of the modern period, John Bourke (1891) painstakingly documents testimonies of anthropologists and ethnographers, from western Europe and the Americas, that give pause to what they perceive to be the 'uncivilized' urinary positions assumed by the immoral throughout history (in and out of Europe, the Americas, and Britain): 'The apache men in micturating always squat down, while the women, on the contrary always stand up ... [quoted in 'Opera,' edited by James Dimock, and published under the direction of the Master of the Rolls, London 1867, 172] ... The author has seen an Italian woman of the lower class urinating in this manner [standing] in the street near San Pietro in Vinculis, Rome, in open daylight, in 1883 ... French women were to be seen in the streets of Paris urinating while standing over gutters [quoted by Mr W.W. Rockhill] ... Among the Turks, it is an heresy, to p---s standing [quoted in Harington, 'Ajax,' 43] ... The Egyptian women stand up when they make water, but the men sit down [quoted in Herodotus, 'Euterpe,' 35] ... the Australian men squatted while urinating; the women generally stood erect [quoted in Mr Carl Lumholtz, 'Among Cannibals,' 1889] ... Old women in Switzerland urinate standing, especially in cold weather [quoted by Rev. Mr Chatelain, a Protestant missionary from Switzerland in Angola, West Africa]' (148–50).

There were also poems written by colonial travellers from England in the late nineteenth century about what appears to them to have been a gender and sexual role reversal:
Behold the strutting Amazonian whore!
She stands in guard, with her right foot before:
Her coat tucked up, and all her motions just,

She stamps, and then cries, 'Hah!' at every thrust.
   But laugh to see her, tired from many a bout,
   Call for the pot, and like a man piss out
   (Juvenal, Satire VI., Dryden's translation, quoted in Bourke 1891, 149).
15 Inglis and Holmes (2000) suggest that it is 'productive to conceptualize the genesis of modernity as in part involving attempts to regulate increasingly the defecatory capacities of human bodies in line with certain normative projects of what is "acceptable" behaviour' (224). They note that modernity is very much about curtailing the time and place of defecation to coincide with the rhythms of the capitalist workday. Toiletry habitus is also about 'faecal invisibility' (ibid., 226). Elimination is unproductive and impolite. It is also noteworthy that families and schools were intent on toilet training children in the modern practice of 'defecating in legitimate locales, and of deferring the time of excretion until a period when the bowels could be exercised in private' (ibid., 234). Hospitals, prisons, asylums, and virtually all institutional settings sought to routinize and privatize toiletry practices.

5. Touching Gender: Abjection and the Hygienic Imagination

1 Rohan, who is transmasculine and butch, says that the 'ultimate fear is you're going to sit on that toilet seat that has AIDS and you're going to take it home to your family.' Chloe, who is a non-trans queer femme, confirms that 'gay bodies – especially gay male bodies – are sometimes perceived to be, in dominant culture, HIV positive.'
2 Olfactory intolerance intersects with what Corbin (1986) refers to as 'the rise of narcissism, the retreat into private space, the destruction of primitive comfort, the intolerance of promiscuity' (232).
3 The interview data show that those who struggle with housing and employment, and those who do custodial work, are culturally aligned with faeces and 'dirt.' As Zoe notes, 'If somebody came into the washroom that was really dirty...or ... [who] was a homeless person that lived on the street ... I'm guessing that people would think of those people as dirty.' Chloe similarly comments, 'I think that people who ... probably are having as ... hard a time as queer people are ... people who are homeless ... perceived as dirty or unclean or things like that, because I think that, um, there's not as much openness to sharing skin contact on a toilet seat with somebody [who is homeless], which is a fairly personal act.' Numerous references are also made to those employed to clean bathrooms, how they are either invisible or seen to be 'unclean.' For a detailed discussion of

how the under-housed are refused access to public toilets, see Sandra Wachholz, 'Hate Crimes against the Homeless' (2005).

4 In her discussion of cleanliness, gender, race, and social difference, Adeline Masquelier writes that 'Unwashed hands, greasy clothes, offensive smells, grime on the skin all entered into complex judgments about not only the social position of the "dirty" person but also his or her moral worth' (2005, 6). Donald Edwards and colleagues (2002) found that women and visible minorities were significantly more likely to wash their hands than white men. People of colour were also found to wash their hands far more frequently than those who are white. This is because white, heterosexual men who are not trans are the least likely to be seen as 'dirty' and socially abject.

5 Richard Dyer (1997) argues that the 'theme of whiteness and death takes many forms. Whites often seem to have a special relation with death, to yearn for it but also to bring it to others. Death may be conceived of as something devoutly to be wished but also as terrifying' (208). Whiteness seems to signify life (light, liberty, and prosperity), but is actually a token of death. In film and popular culture the 'terror of whiteness, of being without life, of causing death, is both vividly conveyed and disowned' (ibid., 210).

6 For a discussion of Victorian designs and ornamentation on early lavatories, see Lucinda Lambton, *Temples of Convenience and Chambers of Delight* (2007); and Kit Wedd, *The Victorian Bathroom Catalogue* (1996).

7 See Guy Hocquenghem (1978) for a good discussion of how the learning of personal hygiene is metonymically associated with private property, class, and capitalist accumulation.

8 See Andil Gosine (2009) for a discussion of the racism endemic to the criminalization of sodomy.

9 The white, sanitized bathroom aesthetic is colonial and imperialist by design. 'Ideas about cleanliness condensed a range of bourgeois values, among them monogamy (clean sex), capitalism (clean profit), Christianity (being cleansed of sin), class distinction, rationality, racial purity' (Boddy 2005, 169). The 'semiotics of boundary maintenance' (Masquelier 2005, 7), inscribed by soap and rituals of personal hygiene and purification, are secured by the establishment of difference, distance, and identity-based exclusions.

10 See Iris Young (1990) for a discussion of how 'socially abjected groups' (145) are constructed as ugly.

11 In an interesting sociological discussion of faecal habitus, Weinberg and Williams (2005) found that 'vigilance concerning breaches of body boundaries [farting, flatulence, and faecal discharge in toilets] was of greater

concern for heterosexual women whose body image was mediated by cultural notions of "feminine" demeanor' (332). Significant gender and sexual differences were noted in the Weinberg and Williams study. For example, while 'heterosexual men can show their power through indifference to the fecal habitus [aligned with femininity and domesticity] ... non-heterosexual men can show resistance to masculinist hegemony by their greater respect for body boundaries' (ibid., 332). It is also worth noting that by keeping the anus clean, gay men mark the anus as a sexual zone. By completely removing faecal matter from the anal cavity through relatively stringent cleaning rituals, men who have sex with men preserve the anus as a sexual orifice.

12 It may seem odd to regard urine and excrement as anything but disgusting. But the unprecedented loathing of so-called body wastes is relatively modern. John G. Bourke (1891) writes, in the now classic *Scatalogic Rites of All Nations*, that modern European sewer systems and latrines were not the result of innovations in human hygiene and the control of germs and disease. Nor were they built to keep urine and faecal matter from fouling the streets of cities like London and Paris. Instead, he suggests, latrines were a means to contain the otherwise potent and sensorial powers of body fluids: 'Enough testimony has been accumulated to convince the most skeptical that the belief was once widely diffused of the power possessed by sorcerers, *et id omne genus*, over the unfortunate wretches whose excreta, solid or liquid, fell into their hands; terror may, therefore, have been the impelling motive for scattering, secreting, or preserving in suitable receptacles the alvine dejections of a community' (Bourke 1891, 134).

Bourke unearths anthropological evidence to show that women bathed in urine to induce conception and drank the fluid to ease the pains of pregnancy, and that the 'urine of eunuchs was considered to be "highly beneficial as a promoter of fruitfulness in females" ' (ibid., 233). His primary sources suggest that urine and faeces were aphrodisiacal, mysterious, and bewitching to people in *both* the Western and non-Western worlds.

13 Weinberg and Williams (2005) suggest that for heterosexual men 'bodily grossness may be valued for its opposition to the manners that femininity is thought to imply ... Some men may adopt this form of embodiment as an expression of their power over women as they deliberately breach this habitus' (317).

14 There is interesting research on the history, cultural politics, and symbolism of shit in modern cultures. See, for example, Begona Aretxaga, 'Dirty Protest' (1995), for a fascinating discussion of the excremental protest by

the Irish National Liberation Army against the British government in a Northern Ireland prison. In her discussion of the 'powers of ordure,' literature, satire, and the metaphor of human excrement, Kelly Anspaugh argues that the anus and the mess it makes have been subject to repression in ways even the genitals and sexual fluids (vaginal fluid and semen, for example) have not. 'Whereas sex has for centuries, at least in the Western world, been sublimated into romantic love, excrement, despite the efforts of a handful of alchemists and the Marquis de Sade, has remained irredeemably base matter' (Anspaugh 1994, 4). In his study of excrement in literature, John R. Clark (1974) writes 'For what society normally considers "low" and "sordid," rhypological and rhyparographical, is more frequently excretory than sexual. Many a man is willing to boast of his sexual prowess and caprice, but is distinctly unwilling to tender public pronouncements about the size of his faeces, the shape of his intestinal disorders, or the stature of his last bout with diarrhea. And a man might be willing to look into another man's sex life, but not into his stool' (43).

15 For a discussion of toilet humour, see Sigmund Freud, *Jokes and Their Relation to the Unconscious* (1975[1905]) and read Alfred Jarry's play *Ubu Roi* (1896), which scandalized late-nineteenth-century Parisian audiences at the Théâtré de l'Oeuvre by its satiric use and flaunting of the scatological.

16 Lynn Sacco (2002) convincingly argues that, in America at the turn of the twentieth century, fears about contracting sexually transmitted infections (such as gonorrhoea and syphilis) from toilet seats were less about evidence-based information about disease transmission than about denying incest in middle- and upper-class family homes. White girls from well-to-do families were said to have contracted such diseases from toilet seats. By contrast, girls with sexually transmitted infections from working-class, immigrant, African-American, and poor families were 'victims' of incest.

17 Kristeva notes that human excrement is the 'most striking example of the interference of the organic within the social' (1982, 75). Organic waste interferes with subject demarcations enabled by language in the symbolic circuit. Binary gender regimes are dependent upon clear subject positions. As literary and art critics argue (Anspaugh 1994, 1995; Canning 1993; Clark 1974; Esty 1999), faecal matter denotes radical ambiguity, disorder, horror, alienation, mortality, and 'matter out of place' (Douglas 1966). Ordure is also used to link 'modern' sanitary engineering with the aesthetics of colonial whiteness. While metaphorics of shit figure prominently in postcolonial writing (Beckett 1970; Joyce 1961 [1922], 1976 [1939], 1976 [1916]; Soyinka 1978 [1965]; Armah 1968), less work has been done on the use of faecal imagery in literature to conjure up images of gender disorder.

18. See Michel Foucault (1978) for a discussion of pastoral power in the secular age.
19. For a discussion of the fluid-bounded body in art and literature, see Claudia Benthien, *Skin* (2002); and Steven Connor, *The Book of Skin* (2004).
20. As C. Jacob Hale (2009) notes in his discussion of transsexuality, voice, and agency, one's insertion into language, as a social subject, demands 'gendered stability both over time and at any given time that some of us lack' (53). For those who undergo gender transitions, there are no gender pronouns or linguistic devices to denote male or female histories that may be at odds with one's present gender identity. If gender pronouns are the toilets of language, as Bobby Noble (2006) notes in his discussion of gender incoherence and trans-masculinities, then it should not be surprising that a cissexist grammar is built into the architecture and designs of toilets.
21. These observations are substantiated by Weinberg and Williams (2005), who observe that urinary and faecal discharge is particularly embarrassing to women: 'heterosexual women were likely to anticipate more deviant labeling, more disgust from other people, and more negative effects on relationships and to report more worry about the loss of attractiveness ... they were also more likely to report that they would not use a public toilet when others were in the facility' (330).
22. Alain Corbin (1986) notes that it was not always the case that excreta and excrement were noxious and offensive to one's olfactory tastes. In nineteenth-century Europe, a growing intolerance for abject body fluids manifested itself in deodorizing and sanitation campaigns focused upon public spaces. 'It was as if thresholds of tolerance had been abruptly lowered; and that happened well before industrial pollution accumulated in urban space' (56). Frederick Engels (1844) traced the building of city sewers not to a societal worry about stink or disease control but to a white, bourgeois disdain for the visibility of working-class and poverty-stricken people. As Steve Pile (1996) writes, 'Similar metonymic chains of associations were built up around other low-Others: thus, slums were linked to dirt, dirt to sewage, sewage to disease, disease to moral degradation, and moral degradation to the slum-dweller or to the prostitute' (180).
23. David Inglis (2002) suggests that 'one of the myriad ways in which dominant groups in colonial and post-colonial contexts have sought to create and recreate the cultural and biological "inferiority" of subaltern groups and classes is through the means of representing the latter in terms that refer to the human body's capacity to create faecal waste' (208). In *Civilization and Its Discontents*, Sigmund Freud argues that beauty, order, and cleanliness are indicators of modern civilization. 'The urge for cleanliness arises from

the wish to get rid of excrement, which has become repugnant to the senses' (Freud 2002 [1930], 42). Anal eroticism, 'childish' indifference to the smell of faeces, and 'primitivism' are all antithetical to a white European ideal of civilization. This ideal, according to Freud, involves the development of the following character traits (all of which derive, in part, from toilet training): order, industry, ambition, competition, sublimated or repressed homosexuality, cleanliness, thriftiness, parsimony, and distance from nature.

24 In Freudian psychoanalysis the penis is symbolized by faeces. Sigmund Freud (1960 [1917]) gives voice to what he observes to be a repressed element of anal eroticism in the child's early, instinctual development: 'The faecal mass, or as one patient called it, the faecal "stick," represents as it were the first penis, and the stimulated mucous membrane of the rectum represents that of the vagina ... during the pregenital phase they had already developed in phantasy and in perverse play an organization analogous to the genital one, in which penis and vagina were represented by the faecal stick and the rectum' (300).

Freud also observes that young children, confused about where babies come from, will sometimes assume that infants are born through the rectum. By metonymic equation there is a link made between faeces, penis, and baby. Sexual difference is confused by the slippages between these three objects. While we cannot infer personality types, genders, or desires from an essential relation to faecal discharge (or to the anus), it is worth noting that anal eroticism connected to the evacuation and exhibition of stool is culturally associated with adolescent rebellion. Freud writes, 'Anal eroticism finds a narcissistic application in the production of defiance, which constitutes an important reaction on the part of the ego against demands made by other people' (ibid., 301).

25 While it may be difficult to think about human excrement as anything but unsexy in the present day, it figures prominently in histories of heterosexual love and was, in fact, regarded as an aphrodisiac. In his discussion of courtship and marriage, Bourke (1891) shares folklore of European history: 'Love-sick maidens in France stand accused of making as a philter a cake into whose composition entered "nameless ingredients," which confection, being eaten by the refractory lover, soon caused a revival of his waning affections' (216). Although use of such philtres was punishable by death, it demonstrates that human excrement was seen to have aphrodisiac qualities. Bourke also provides an account of a practice in which what we might call faecal-cakes were put on the backside: 'The method of divination by which maidens strove to rekindle the expiring flames of affection in the hearts of husbands and lovers by making cake from dough

kneaded on the woman's posterior ... seems to have held on in England as a game among little girls, in which one lies down on the floor, on her back, rolling backwards and forwards, and repeating the following lines: – *"Cockledy break, mistley cake, When you do that for our sake."* While one of the party so lay down the rest of the party sat round; they lay down and rolled in this manner by turns' (ibid., 221). In her fascinating history of chocolate, Alison Moore (2005) argues that European appetites for cocoa are related to taboos on coprophagia – a taste for human excrement. Chocolate, a gift exchanged by lovers on Valentine's Day, has, 'throughout the late modern era ... been repeatedly associated, both explicitly and symbolically, with excrement' (52). Moulded into the shape of eggs, logs, and other *'lumpf-like forms'* (ibid., 59), chocolate evokes thinly disguised coprophilic fantasies. Noticing that the advent of the modern toilet coincides with the colonial appropriation of cocoa from Latin American countries, Moore argues that chocolate 'functioned as a symbol of the erotic, the infantile, and the feminine aspects attributed to primitivity and which were cast out as waste matter in the masculine, adult work of civilized society and capitalist economic order' (ibid., 52–3).

26 For a discussion of how faeces are artistically smeared on bathroom walls, see A. Dundas, 'Here I Sit' (1966).
27 See Linda Williams (1989) for a discussion of ejaculation scenes in hard-core pornography. See also Thomas (2008) for a discussion of how a public display of male ejaculate sometimes represents the feminine because of its visibility; the 'male matter appears to be feminine, is feminized upon appearance' (4).
28 It should also be noted that people miss the toilet bowl and spread faeces because of trouble with psychiatric medications, mental health, and/or incontinence.
29 For a discussion of anal pleasure and its discipline in the United States, see Jack Morin, *Anal Pleasure and Health* (1998).
30 Allen Chun (2002) writes about what he calls the development of the 'supermodern Japanese bidet-toilet' and the Western automated technologies and sensibilities it relies upon. The focus upon cleanliness, sanitation, and the masking of bodily sounds (primarily in the 'women's' toilet) is, he suggests, about the 'filth associated with the traditional Japanese squat toilet' (153). The racialization and concurrent degradation of the squatting toilet can also be seen in the globalization of the Western-style flush toilet.
31 I do not mean to shun toilet technologies that enhance accessibility for persons with physical disabilities. Such technologies are not objectifying so much as they invite touch and a yielding to one's environment.

6. Sexing Gender: The Homoerotics of the Water Closet

1 See Nikki Sullivan (2003) for a critical discussion of sexology and its relation to queer theory.
2 Referring to hedonistic and unproductive sex, Edelman writes, 'If, however, there is *no baby* and, in consequence, *no future*, then the blame must fall on the fatal lure of sterile, narcissistic enjoyments understood as inherently destructive of meaning and therefore as responsible for the undoing of social organization, collective reality, and, inevitably, life itself' (2004, 13).
3 See Judith Butler (1993) for a discussion of the lesbian phallus.
4 Leo Bersani (1986) writes that the anus, 'like the mind expels from the body substances which the body both produces and treats like waste. Thought, far from providing a guarantee of being in this radically non-Cartesian world, is the excrement of being' (9). In other words, there is a close metonymic relationship between forgetting, the anus, and the toilet.
5 For a discussion of gay male public sex cultures, tearooms, and cottages, see Robert Aldrich, 'Homosexuality and the City: An Historical Overview,' *Urban Studies* 41.9 (2004): 1719–37; Don Bapst, 'Glory Holes and the Men Who Use Them' (2001); Lauren Berlant and Michael Warner, 'Sex in Public,' *Critical Inquiry* 24.2 (1998): 547–66; Edward William Delph, *The Silent Community: Public Homosexual Encounters* (London: Sage Publications, 1978); Frederick J. Desroches, 'Tearoom Trade: A Research Update,' *Qualitative Sociology* 13.1 (1990): 39–61; Lee Edelman, 'Capital Offenses: Sodomy in the Seat of American Government,' in *Homographies*, 129–88 (1994); Lee Edelman, 'Tearoom and Sympathy; or, The Epistemology of the Water Closet,' in *Homographies*, 148–70 (1994); Paul Flowers, Claire Marriott, and Graham Hart, 'The Bars, the Bogs, and the Bushes: The Impact of Locale on Sexual Cultures,' *Culture, Health and Sexuality* 2.1 (2000): 69–86; John Hollister, 'Beyond the Interaction Membrane: Laud Humphrey's Tearoom Tradeoff,' *International Journal of Sociology and Social Policy* 24.3–5 (2004): 72–94; Matt Houlbrook, 'The Private World of Public Urinals' (2000); Laud Humphreys, *Tearoom Trade: Impersonal Sex in Public Places* (New York: Aldine de Gruyter, 1975); Maurice van Lieshout, 'Leather Nights in the Woods: Homosexual Encounters in a Dutch Highway Rest Area,' *Journal of Homosexuality* 29.1 (1995): 19–39; William L. Leap, *Public Sex/Gay Space* (New York: Columbia University Press (1999); Sonia Magni and Vasu Reddy, 'Performative Queer Identities' (2007); Jeffrey Merrick, 'Sodomites and Police in Paris, 1715,' *Journal of Homosexuality* 42.3 (2002): 103–28; Peter M. Nardi, 'The Breastplate of Righteousness: Twenty-five Years after Laud Humphrey's *Tearoom Trade: Impersonal Sex in Public Places*,' *Journal of*

Homosexuality 30.2 (1995): 1–10; John Potvin, 'Vapour and Steam: The Victorian Turkish Bath, Homosexual Health, and Male Bodies on Display,' *Journal of Design History* 18.4 (2005): 319–33; Richard Tewksbury, 'The Intellectual Legacy of Laud Humphreys' (2004); Richard Tewksbury, 'Cruising for Sex in Public Places: The Structure and Language of Men's Hidden, Erotic Worlds,' *Deviant Behavior: An Interdisciplinary Journal* 17 (1996): 1–19.

6 The installation of short doors in lavatories is not a new development. Michel Foucault (1978) argued that the modern latrine was originally designed with sex in mind. Focusing on the panoptic designs of school washrooms, he writes that 'latrines had been installed with half-doors, so that the supervisor on duty could see the head and legs of the pupils, and also with side walls sufficiently high "that those inside cannot see one another" ' (173).

7 As Clara Greed (1995) notes, toilets in Britain are closed at six p.m. to prevent public sex. While lobbying for public restroom accessibility for women at night, Greed found that resistance to her efforts was fuelled by 'Fears of prostitution, theft, underage sex, drug trafficking, graffiti, and vandalism' (576). She also suggests that 'Male toilets used for cottaging ... would be more likely to get police surveillance than be closed' (ibid.).

8 For a historical discussion of gay male public sex cultures and policing in Toronto, see Gary Kinsman, *The Regulation of Desire: Sexuality in Canada* (Montreal: Black Rose Books, 1987); Steven Maynard, 'On the Case of the Case' (1998); and Steven Maynard, 'Through a Hole in the Lavatory Wall' (1994).

9 Interviewees confirm that the removal of doors on stalls is a barrier to access. Brant, who is a non-trans queer man, explains that at the University of Toronto, 'in one of the bathrooms, there was so much gay sex going on that they actually took off the stall doors ... which was amazingly frustrating for me. I walked in and it's like, "fuck!" [... because I can't use them].'

10 In present-day Toronto, Ivan, who is a non-trans queer man, notices how mirrors are positioned in bathrooms in such a way that they turn patrons into spectacles – so that you can 'see someone in all directions ... [with] these new forms of architecture where there's no doors now, so people who enter glide through silently, it's like a labyrinth ... there's [no] privacy. Dark rooms are also constructed that way ... [but] there's a light trap ... but there's no physical barrier. Now of course the idea there is also to be very free, so that people in wheelchairs and others with mobility [constraints] can have access. So it has this dual function for increasing access for those who are disabled, but also it serves as a form of ... passive social control around so-called unauthorized sexual activities that happen between men.'

11 Those who use the 'women's' room talk extensively about heterosexist advertising. Sugar, who is a non-trans queer femme, says she sees ads for 'yeast infections ... [everything from] sexuality to pregnancy, to ... the morning-after pill or plan B or whatever.' Kew, who is genderqueer and a sex educator, also comments on the gendered advertisements for birth control and medications for sexually transmitted infections in public toilets: '[There are] ... definitely birth-control pills and ... herpes medications ... Morning-after pills ... we see [endorsements for the birth-control pill] Alesse a lot in the subway [toilets] ... But the morning-after pills we don't see in the subways. I think it's a little bit more risqué because it is a little more explicit because [the] sex is more recent ... So ... definitely it's very gendered targeted advertisement.'

Callum, who is a trans guy, similarly notes that corporate advertisements in toilets on Canadian university campuses 'seem very gendered. In women's washrooms it's birth-control pills, yeast-infection meds ... [and] other thoroughly feminine coded ads.' He also notices the extensive focus upon 'feminine hygiene, or hygiene [products in general].' There are also advertisements for on-line heterosexual dating services. Eric Prete, who is a trans man, complains that he always sees 'anti-pregnancy ad[s] ... which [are] ... pretty offensive in and of themselves ... There's the assumption that queers don't exist in these spaces.' Neil, who is transmasculine, laments that 'If you are trying to pee and there's this thing staring at you in the face telling you to take birth control, that's the kind [of thing] that I am really annoyed with, especially in [the] student union building [at his Ontario university], where you are sitting down trying to take a leak and there's an ad right across from you and it's like "Ask your doctor about ... birth control." It's just awkward ... Like with the birth-control ads, automatically it assumes the person identifies as female and identifies as heterosexual ... I mean, if you are sitting there and you don't identify as either, you just feel kind of lost.'

12 The only pseudo gay-positive advertisements seen in the lavatory seem to be sponsored by financial institutions. As observed by Laura, who is queer, the advertisements target white, gay couples interested in home ownership and joint credit cards: 'I've actually been surprised that there are a number of ads that feature what looks like two women that may be a couple, or what looks like two men who may be a couple, who are interested in a mortgage or a vacation credit card.

13 Brant offers the following direct quotations from an Internet cruising site to encapsulate his observations about gay, white, male racism as manifested in hostility to cruising strategies employed by Asian men:

(1) To the Filipino guy: if you want to cruise, that's fine, but learn some etiquette about how to be discreet. First, don't slide gay advertisements under each and every occupied cubicle especially when they are occupied and you have no idea whether the other guy is there to urinate/defecate. Two, don't stand in front of a stall after you know the other guy isn't into you. You are cock-blocking ... If he's ignoring you, he's not interested.

(2) The Chinese man: washing your hands every five minutes is great, and I applaud you on cleanliness, but don't loiter by picking your eyebrows and combing your hair over and over again. Decide what you want to pursue and go for it. But don't hang around looking obviously in the hopes that you will see cock. Guys like to show their cock to guys who are willing to suck it, show their cock, or play – not to guys who just stand around acting weird.

(3) Place was busy at 12:30, however, that Asian cock blocker who washes his hands over and over kills the action. Honestly, they are clean, you don't need to wash them again as you aren't touching your cock/ass nor anyone else's. You aren't even sitting on a toilet. I don't know why when I told you to stop washing your hands over and over, you sat there with a dumb ass I don't speak English look. Either learn English and learn how to cruise, or leave and masturbate at home.

Given the racist commentary, further investigation of the racial cultural politics of gay male cruising cultures is clearly needed.

14 In an important discussion of the sexual and racial politics of environmental movements, Andil Gosine (2009) argues that heterosexual sex between non-white people is regarded as dangerous and threatening to the global ecosystem because it is imagined by white people to be the cause of 'overpopulation.' While gay white men are threatening to heteronationalist body politics because they (allegedly) do not reproduce, non-white men who have heterosexual sex, families, and children are imagined to be a threat to global environmental sustainability. It should be noted that colonial and eugenic discourses about race are deeply embedded in white reproductive futurity.

15 Conversely, white cissexual men are sometimes sexually intimidated by images of black male hetero-virility. Richard Dyer (1997) also notes that it is a 'specifically white, aghast perception of the unstoppable breeding of non-whites, the deep-seated suspicion that non-whites are better at sex and reproduction than are whites, that, indeed, to be truly white and reproductively efficient are mutually incompatible and that, as a result, whites are going to be swamped and engulfed by the non-white multitudes' (216).

16 For a discussion of how Asians are associated with the anus in American pornography, see Richard Fung, 'Looking for My Penis' (1991).
17 It is worth considering how the toilet stall may operate as a late-modern confessional where one waits for a written reply from an anonymous correspondent – one who can answer a question (or comment upon a sexual act) that is unsuitable for more direct conversations between family and friends.
18 It is interesting to note that in early-nineteenth-century Rome, authorities attempted to curtail 'obscene poetry' 'by consecrating the walls so exposed with the picture of a deity or some other hallowed emblem, and by denouncing the wrath of heaven against those who should be impious enough to pollute what it was their duty to reverence ... The snake, it is well known, was reckoned among the gods of the heathens' (Bourke 1891, 136). Sexually graphic and denigrating images of people now drawn on cubicle partitions mark a dramatic contrast to the images of Roman goddesses, such as Cloacina, that were once painted on latrine walls to curtail graffiti.
19 For a discussion of gender and graffiti, see Bob Alexander, 'Male and Female Rest Room Graffiti,' *Maledica* 2.1 (1978): 42–59; Lynn Bartholme and Philip Snyder, 'Is It Philosophy or Pornography? Graffiti at the Dinosaur Bar-B-Que,' *Journal of American Culture* 27.1 (2004): 86–98; Patricia Cooper, 'Cherished Classifications' (1999); Jane M. Gadsby, 'To My Fellow Sisters: Discourse on the Washroom Walls (Latrinalia),' *Canadian Folklore* 18 (1996): 27–48; James A. Green, 'The Writing on the Stall: Gender and Graffiti,' *Journal of Language and Social Psychology* 22.3 (2003): 282–96; Judith Halberstam, 'Techno-Homo' (1997b); and George E. Schreer, 'Private Restroom Graffiti: An Analysis of Controversial Social Issues on Two College Campuses,' *Psychological Reports* 81.3 (1997): 1067–74.
20 For a discussion of racist graffiti, see Samuel Gyasi Obeng, 'Speaking the Unspeakable' (2000).
21 In *The Anatomy of Prejudices,* Elisabeth Young-Bruehl contends that homophobia is about preserving gender sameness. Homosexuals are thought to 'confuse the fact of anatomical difference, they upset the world of sexual identity concepts and boundaries' (1996, 36). There is, she tells us, overlap between homo- and transphobias. This is not to say that there are not quantifiable differences in the deployment of homo- or transphobias but that they are not always distinguishable. Young-Bruehl observes that 'Most attacks on homosexuals are conducted by males in their late teens and early twenties, just when they find assertions of their own masculinity most necessary for identity consolidation' (ibid., 239). A driving force behind cissexual male homophobia may be a reluctance to avow a feminine

side to the self. Young-Bruehl argues that 'men are contemptuous of feminine gay men and anxious about being approached or raped by masculine gay men – although both feelings may be registrations of the same anxiety about feminization' (ibid., 150).

22 The queer or sinthomosexual, in Edelman's economy, is a child-negating and future-denying force. The child embodies the 'telos of the social order and come[s] to be seen as the one for whom that order is held in perpetual trust' (Edelman 2004, 11).

23 For critical discussions of paedophilic panics, see Sheila Cavanagh, *Sexing the Teacher* (2007); Dean Durber, 'The Paedophile and "I",' *Media International Australia* 127 (2008): 57–70; James Kincaid, *Erotic Innocence: The Culture of Child Molesting* (Durham and London: Duke University Press, 1998); Richard D. Mohr, 'The Pedophilia of Everyday Life,' in *Curiouser: On the Queerness of Children*, ed. Steven Bruhm and Natasha Hurley (Minneapolis: University of Minnesota Press, 2004).

24 For a critical discussion of the endangered child, see Marjorie Heins, *Not in Front of the Children: 'Indecency,' Censorship, and the Innocence of Youth* (New York: Hill and Wang, 2001); and Judith Levine, *Harmful to Minors: The Perils of Protecting Children from Sex* (New York: Thunder's Mouth Press, 2003).

25 The restroom on the third floor of the Toronto Eaton Centre is a well-known cruising area for gay men, and there has been public controversy leading to security and police surveillance.

26 For a discussion of how the lavatory provided an opportunity for lower-middle-class and working-class men in Toronto to have sex with men in the early twentieth century, see Steven Maynard, 'Through a Hole in the Lavatory Wall' (1994).

Conclusion

1 See Wright (1960) for a discussion of how the Victorian plumber escalated in social rank when in 1871 the Prince of Wales (after recovering from typhoid) pronounced that plumbing was a noble profession.

2 Some interviewees believe that public bathrooms should be queer and sex-positive. JB says that we need to 'get away from thinking that washrooms are this breeding ground for sexual deviance.' One interviewee who is Hong Kong Canadian and queer insists that there should 'definitely [be] condoms ... and ... other types of contraception [in the bathroom] ... [and there should be] gloves or ... lubricant.' Haley says in jest that the 'door handle would [in an ideal world] have levers ... like, a funny dildo handle

[for public sex] on the bathroom door ... rather than one of those straight, metal, cold bars. Maybe have sex toys available in bathrooms ... And free condoms ... with any kind of sex toy, not just dildos, but like, cock rings, nipple clamps ... blindfolds and floggers and a whole array [of sex toys].' Haley also says that toilets should have 'little benches and little sofas like in Buchanan Tower so that if people wanted to have sex they could have sex.'

3 In his discussion of the pneumatic bathroom, Marco Frascari (1997) writes that the lavatory at home is the 'ideal place for fostering a beautiful life' (165). He further suggests that the bathroom is a numinous place: 'During everyday use of the bathroom, hierophanies can take place. A hierophany separates the thing that manifests the sacred from everything else around it ... The same thing can be said about the bathroom and its objects, since they can draw together meanings of initiation, love (both profane and sacred), beauty and ugliness, mutation and transformation, and birth' (ibid., 167).

4 In her cultural history of skin, Claudia Benthien (2002) writes that, beginning in the eighteenth century, a new body metaphor was adopted: that of the body as 'house.' We live 'in' our bodies, are 'imprisoned' by them, 'encased' or 'enveloped' by them. Skins are seen as walls and the body as surface, as the 'place where identity is formed and assigned' (1). Orifices are like doors and portholes to the self. Sensations are like windows to the soul. What is unique about the 'house model of sensory perception' (ibid., 28) is its reliance upon modern-day optics.

5 See, for example, Christopher Butterfield's *Stall*, which premiered at the Voice Over Mind festival in Vancouver, British Columbia, May 2010, in the 'women's' washroom at the Chan Centre for the Performing Arts.

Glossary

1 See http://www.dcatsinfo.org (accessed 6 March 2009).
2 For a discussion of two-spirited people, see Sue-Ellen Jacobs, Wesley Thomas, and Sabine Lang, *Two-Spirit People: Native American Gender Identity, Sexuality, and Spirituality* (Chicago: University of Illinois Press, 1997).

Bibliography

Aitken, S. 2001. *Geographies of Young People: The Morally Contested Spaces of Identity*. London: Routledge.

Aitken, S., and T. Herman. 1997. 'Gender, Power and Crib Geography: Transitional Spaces and Potential Places.' *Gender, Place and Culture* 4: 63–88.

Anderson, Richard. 2008. 'Gender, Class and Bodily Functions in the Urban Past: A Preliminary Historical Geography of Toronto's Public Lavatories.' Unpublished paper.

Anderson, Warwick. 1995. 'Excremental Colonialism: Public Health and the Poetics of Pollution.' *Critical Inquiry* 21: 640–69.

– 2002. 'Going through the Motions: American Public Health and Colonial "Mimicry." ' *American Literary History* 14.4: 686–719.

Anspaugh, Kelly. 1994. 'Powers of Ordure: James Joyce and the Excremental Vision(s).' *Mosaic: A Journal for the Interdisciplinary Study of Literature* 27.1: 73–100.

– 1995. 'Ulysses upon Ajax? Joyce, Harington, and the Question of "Cloacal Imperialism." ' *South Atlantic Review* 60.2: 11–29.

Anzieu, Didier. 1989. *The Skin Ego*. Translated by Chris Turner. New Haven and London: Yale University Press.

Aretxaga, Begona. 1995. 'Dirty Protest: Symbolic Overdetermination and Gender in Northern Ireland Ethnic Violence.' *Ethos* 23.2: 123–48.

Armah, Ayi Kwei. 1968. *The Beautiful Ones Are Not Yet Born*. Portsmouth: Heinemann.

Bapst, Don. 2001. 'Glory Holes and the Men Who Use Them.' *Journal of Homosexuality* 41.1: 89–102.

Bataille, Georges. 1957. *Erotism: Death and Sensuality*. San Francisco: City Lights Books.

Beckett, Samuel. 1970. *First Love and Other Shorts*. New York: Grove Press.

Benjamin, Walter. 1978. 'On the Mimetic Faculty.' In *Reflections*, trans. Edmund Jephcott, 333–6. New York: Harcourt.

Benthien, Claudia. 2002. *Skin: On the Cultural Border between Self and the World*. New York: Columbia University Press.

Berlant, Lauren, and Michael Warner. 2002. 'Sex in Public.' In *Publics and Counterpublics*, ed. Michael Warner, 187–208. New York: Zone Books.

Bersani, Leo. 1986. *The Freudian Body: Psychoanalysis and Art*. New York: Columbia University Press.

– 2001. 'Genital Chastity.' In *Homosexuality and Psychoanalysis*, ed. Tim Dean and Christopher Lane, 351–66. Chicago and London: University of Chicago Press.

Bess, Philip. 1997. 'Democracy's Private Places.' *First Things: A Monthly Journal of Religion and Public Life* 76: 16–18.

Bhabha, Homi. 1994. *The Location of Culture*. London: Routledge.

Blair, Monroe. 2000. *Ceramic Water Closets*. Princes Risborough: Shire Publications, Ltd.

Blesser, Barry, and Linda-Ruth Salter. 2007. *Spaces Speak, Are You Listening? Experiencing Aural Architecture*. Cambridge, MA: MIT Press.

Boddy, Janice. 2005. 'Purity and Conquest in the Anglo–Egyptian Sudan.' In *Dirt, Undress, and Difference: Critical Perspectives on the Body's Surface*, ed. Adeline Masquelier. Bloomington and Indianapolis: Indiana University Press.

Bourke, John G. 1891. *Scatalogic Rites of All Nations*. Washington: W.H. Lowdermilk and Co.

Boyd, Nan Alamilla. 2006. 'Bodies in Motion: Lesbian and Transsexual Histories.' In *The Transgender Studies Reader*, ed. Susan Stryker and Stephen Whittle, 420–33. New York: Routledge.

Braham, William W. 1997. 'Siegfried Giedion and the Fascination of the Tub.' In *Plumbing: Sounding Modern Architecture*, ed. Nadir Lahiji and D.S. Friedman. New York: Princeton Architectural Press.

Brennan, Teresa. 2004. *The Transmission of Affect*. Ithaca and London: Cornell University Press.

Britzman, Deborah. 1998. *Lost Subjects, Contested Objects: Toward a Psychoanalytic Inquiry of Learning*. Albany, NY: State University of New York Press.

Bronfen, Elisabeth. 1998. *The Knotted Subject: Hysteria and Its Discontents*. Princeton, NJ: Princeton University Press.

Browne, Kath. 2004. 'Genderism and the Bathroom Problem: (Re)Materializing Sexed Sites, (Re)Creating Sexed Bodies.' *Gender, Place and Culture* 11.3: 331–46.

Butler, Judith. 1990. *Gender Trouble: Feminism and the Subversion of Identity*. New York: Routledge.

- 1993. *Bodies That Matter: On the Discursive Limits of 'Sex.'* New York: Routledge.
- 1997a. *Excitable Speech: A Politics of the Performative.* New York: Routledge.
- 1997b. *The Psychic Life of Power: Theories in Subjection.* Stanford, CA: Stanford University Press.
- 2004. *Undoing Gender.* New York and London: Routledge.
- 2005. *Giving an Account of Oneself.* New York: Fordham University Press.

Cahill, Spencer E., with William Distler, Cynthia Lachowetz, Andrea Meaney, Robyn Tarallo, and Teena Willard. 1985. 'Meanwhile Backstage: Public Bathrooms and the Interaction Order.' *Urban Life* 14.1: 33–58.

Callard, Felicity. 2003. 'The Taming of Psychoanalysis in Geography.' *Social and Cultural Geography* 4.3: 295–312.

Camfield, William. 1989. *Marcel Duchamp: Fountain.* Houston: Houston University Press.

Canning, Susan M. 1993. 'The Ordure of Anarchy: Scatological Signs of Self and Society in the Art of James Ensor.' *Art Journal* 52.3: 47–54.

Cavanagh, Sheila L. 2003. 'Teacher Transsexuality: The Illusion of Sexual Difference and the Idea of Adolescent Trauma.' *Sexualities: Studies in Culture and Society* 6.3–4: 365–88.

- 2007. *Sexing the Teacher: School Sex Scandals and Queer Pedagogies.* Vancouver: University of British Columbia Press.

Cavanagh, Sheila, and Heather Sykes. 2006. 'Transsexual Bodies at the Olympics: The International Olympic Policy on Transsexual Athletes at the Athens Summer Games.' *Body and Society* 12.3: 75–102.

Chase, Cheryl. 2006. 'Hermaphrodites with Attitude: Mapping the Emergence of Intersex Political Activism.' In *The Transgender Studies Reader*, ed. Susan Stryker and Stephen Whittle, 300–14. New York: Routledge.

Chauncey, George. 1994. *Gay New York: Gender, Urban Culture, and the Making of the Gay Male World, (1890–1940).* New York: Basic Books.

Chess, Simone, Alison Kafer, Jessi Quizar, and Mattie Udora Richardson. 2004. 'Calling All Restroom Revolutionaries!' In *That's Revolting! Queer Strategies for Resisting Assilmilation*, ed. Mattilda AKA Matt Bernstein Sycamore, 189–203. New York: Soft Skull Press.

Chun, Allen. 2002. 'Flushing in the Future: The Supermodern Japanese Toilet in a Changing Domestic Culture.' *Postcolonial Studies* 5.2: 153–70.

Clark, John R. 1974. 'Bowl Games: Satire in the Toilet.' *Modern Language Studies.* 4.2: 43–58.

Cockayne, Emily. 2007. *Hubbub: Filth, Noise and Stench in England, 1600–1770.* New Haven and London: Yale University Press.

Cohen, Deborah, Ute Lehrer, and Andrea Winkler. 2005. 'The Secret Lives of Toilets: A Public Discourse on "Private" Space in the City.' In *Utopia:*

Towards a New Toronto, ed. J. McBride and A. Wilcox, 194–203. Toronto: Coach House Books.
Cohen, William A., and Johnson, Ryan, eds. 2005. *Filth: Dirt, Disgust, and Moral Life*. Minneapolis and London: University of Minnesota Press.
Colman, Penny. 1994. *Toilets, Bathtubs, Sinks, and Sewers: A History of the Bathroom*. New York: Atheneum, Macmillan Publishing Company.
Connor, Steven. 1997a. 'Feel the Noise: Excess, Affect, and the Acoustic.' In *Emotions in Postmodernism*, ed. Gerhard Hoffman, 147–62. Heidelberg: Universitätsverlag C. Winter.
– 1997b. 'The Modern Auditory I.' In *Rewriting the Self: Histories from the Renaissance to the Present*, ed. Roy Porter, 203–23. London: Routledge.
– 2004. *The Book of Skin*. London: Reaktion Books.
Cooper, Annabel, Robin Law, and Jane Malthus. 2000. 'Rooms of Their Own: Public Toilets and Gendered Citizens in a New Zealand City, 1860–1940.' *Gender, Place and Culture* 7.4: 417–33.
Cooper, Patricia. 1999. 'Cherished Classifications: Bathrooms and the Construction of Gender/Race on the Pennsylvania Railroad during World War II.' *Feminist Studies* 25.1: 7–41.
Corbin, Alain. 1986. *The Foul and the Fragrant: Odor and the French Social Imagination*. Cambridge, MA: Harvard University Press.
Costa, Xavier. 1997. 'Ground Level.' In *Plumbing: Sounding Modern Architecture*, ed. Nadir Lahiji and D.S. Friedman, 93–102. New York: Princeton Architectural Press.
Cummings, William. 2000. 'Squat Toilets and Cultural Commensurability: Two Texts Plus Three Photographs I Forgot to Take.' *Journal of Mundane Behavior* 1.3: 261–73.
Davidson, Michael. 2003. 'Phantom Limbs: Film Noir and the Disabled Body.' In *GLQ: A Journal of Lesbian and Gay Studies*, 9.1–2: 57–77.
Davies, Cristyn. 2007. 'Queering the Space of the Public Toilet.' Conference paper presented at *Queer Space: Centers and Peripheries*, University of Technology at Sydney, Australia.
Dean, Tim. 2000. *Beyond Sexuality*. Chicago: University of Chicago Press.
Deleuze, Gilles, and Felix Guattari. 1989. *Anti-Oedipus: Capitalism and Schizophrenia*, trans. Robert Hurley, Mark Seem, and Helen R. Lane. Minneapolis: University of Minnesota Press.
Devor, Aaron H., and Nicholas Matte. 2006. 'ONE Inc. and Reed Erickson: The Uneasy Collaboration of Gay and Trans Activism, 1964–2003.' In *The Transgender Studies Reader*, ed. Susan Stryker and Stephen Whittle, 387–406. New York: Routledge.

Douglas, Mary. 1966. *Purity and Danger: An Analysis of the Concepts of Pollution and Taboo*. Boston: Ark Paperbacks.

Duggan, Lisa. 2003. *The Twilight of Equality? Neoliberalism, Cultural Politics and the Attack on Democracy*. Boston: Beacon Books.

Dundas, A. 1966. 'Here I Sit – A Study of American Latrinalia.' *Kroeber Anthropological Papers* 34: 91–105.

Dutton, Michael, Seth Sanjay, and Leela Gandhi. 2002. 'Editorial: Plumbing the Depths: Toilets, Transparency and Modernity.' *Postcolonial Studies* 5.2: 137–42.

Dwyer, Owen J., and John Paul Jones III. 2000. 'White Socio–Spatial Epistemology.' *Social and Cultural Geography* 1.2: 209–22.

Dyer, Richard. 1997. *White*. New York: Routledge.

Edelman, Lee. 1991. 'Seeing Things: Representation, the Scene of Surveillance, and the Spectacle of Gay Male Sex.' In *Inside/Out: Lesbian Theories, Gay Theories*, ed. Diana Fuss. New York: Routledge.

– 1993. 'Tearooms and Sympathy; or, the Epistemology of the Water Closet.' In *The Lesbian and Gay Studies Reader*, ed. Henry Abelove, Michèle Aina Barale, and David M. Halperin, 553–74. New York: Routledge.

– 1994. *Homographies: Essays in Gay Literary and Cultural Theory*. New York: Routledge.

– 2004. *No Future: Queer Theory and the Death Drive*. Durham and London: Duke University Press.

Edwards, Donald, Elizabeth Monk-Turner, Steve Poorman, Maria Rushing, Stephen Warren, and Jarta Willie. 2002. 'Predictors of Hand-Washing Behavior.' *Social Behavior and Personality* 30.8: 751–6.

Elliot, Patricia. 2001. 'A Psychoanalytic Reading of Transsexual Embodiment.' *Studies in Gender and Sexuality* 2.4: 295–325.

Ellis, Havelock. 1928. *Studies in the Psychology of Sex*. Philadelphia: Davis.

Eng, David L. 2001. *Racial Castration: Managing Masculinity in Asian America*. Durham and London: Duke University Press.

Engels, Frederick. 1844. *The Condition of the Working Class*. Harmondsworth: Panther.

Esty, Joshua D. 1999. 'Excremental Postcolonialism.' *Contemporary Literature* 40.1: 22–46.

Evans, Dylan. 1996. *An Introductory Dictionary of Lacanian Psychoanalysis*. New York: Brunner-Routledge.

Eveleigh, David J. 2002. *Bogs, Baths and Basins: The Story of Domestic Sanitation*. Stroud: Sutton Publishing.

Fanon, Frantz. 1986. *Black Skin, White Mask*. Trans. Charles Lamm Markmann. London: Pluto Press.

Fausto-Sterling, Anne. 2000. *Sexing the Body: Gender Politics and the Construction of Sexuality*. New York: Basic Books.
Felski, Rita. 2006. 'Fin de Siècle, Fin du Sexe: Transsexuality, Postmodernism, and the Death of History.' In *The Transgender Studies Reader*, ed. Susan Stryker and Stephen Whittle, 565–73. New York: Routledge.
Fernbach, Amanda. 2002. *Fantasies of Fetishism: From Decadence to the Post-Human*. New Brunswick, NJ: Rutgers University Press.
Foucault, Michel. 1978. *The History of Sexuality, An Introduction: Volume 1*. New York: Vintage.
– 1979. *Discipline and Punish: The Birth of the Prison*. New York: Vintage.
Frascari, Marco. 1997. 'The Pneumatic Bathroom.' In *Plumbing: Sounding Modern Architecture*, ed. Nadir Lahiji and D.S. Friedman, 163–80. New York: Princeton Architectural Press.
Freud, Sigmund. 1960 [1917]. 'On Transformations of Instinct as Exemplified in Anal Eroticism.' In *On Sexuality: Three Essays on the Theory of Sexuality and Other Works, The Standard Edition, Vol. 7*, ed. and trans. James Strachey, 130–243. London: The Hogarth Press and the Institute of Psycho-Analysis.
– 1960 [1927–1931]. 'Fetishism.' In *The Standard Edition of the Complete Psychological Works, Vol. 21*, ed. and trans. James Strachey, 152–7. London: The Hogarth Press and the Institute of Psycho-Analysis.
– 1961 [1923]. *The Ego and the Id*. In *The Standard Edition of the Complete Psychological Works, Vol. 19*, ed. and trans. James Strachey, 3–59. London: The Hogarth Press and the Institute of Psycho-Analysis.
– 1975 [1905]. *Jokes and Their Relation to the Unconscious*. In *The Standard Edition of the Complete Works of Sigmund Freud, Vol. 8*, trans. James Strachey. London: Vintage Press.
– 2002 [1930]. *Civilization and Its Discontents*. London: Penguin Books.
Friedman, David M. 2001. *A Mind of Its Own: A Cultural History of the Penis*. New York: Free Press.
Fung, Richard. 1991. 'Looking for My Penis: The Eroticized Asian in Gay Porn Video.' In *How Do I Look?*, ed. Bad Object Choices. Seattle: Bay Press.
Fuss, Diana. 1995. *Identification Papers*. New York: Routledge.
Gastelaars, Marja. 1996. 'The Water Closet: Public and Private Meanings.' *Science as Culture* 5: 483–505.
Gilbert, Pamela K. 2005. 'Medical Mapping: The Thames, the Body, and Our Mutual Friend.' In *Filth: Dirt, Disgust, and Modern Life*, ed. William A. Cohen and Ryan Johnson, 78–102. Minneapolis and London: University of Minnesota Press.
Gillison, Gillian. 1980. 'Images of Nature in Gimi Thought.' In *Nature, Culture, and Gender*, ed. Carol P. MacCormack and Marilyn Strathern, 143–73. New York: Cambridge University Press.

Goffman, Erving. 1959. *The Presentation of Self in Everyday Life*. New York: Doubleday.
Gosine, Andil. 2009. 'Non-white Reproduction and Same-Sex Eroticism: Queer Acts against Nature.' In *Green, Pink, and Public: Queering Environmental Politics*, ed. Catriona Mortimer-Sandilands, 227–62. Bloomington and Indianapolis: Indiana University Press.
Greed, Clara. 1995. 'Public Toilet Provision for Women in Britain: An Investigation of Discrimination against Urination.' *Women's Studies International Forum* 18.5–6: 573–84.
Grosz, Elizabeth. 1994. *Volatile Bodies: Toward a Corporeal Feminism*. Bloomington and Indianapolis: Indiana University Press.
Halberstam, Judith. 1997a. 'Bathrooms, Butches, and the Aesthetics of Female Masculinity.' In *Rrose is a Rrose is a Rrose: Gender Performance in Photography*, ed. Jennifer Blessing, 176–89. New York: Guggenheim Museum.
– 1997b. 'Techno-Homo: On Bathrooms, Butches and Sex with Furniture.' In *Processed Lives: Gender and Technology in Everyday Life*, ed. Jennifer Terry and Melodie Calvert. London and New York: Routledge.
– 1998. *Female Masculinity*. Durham and London: Duke University Press.
– 2005. *In a Queer Time and Place: Transgender Bodies, Subcultural Lives*. New York: New York University Press.
Hale, C. Jacob. 2009. 'Tracing a Ghostly Memory in My Throat: Reflections on Ftm Feminist Voice and Agency.' In *You've Changed: Sex Reassignment and Personal Identity*, ed. Laurie Shrage, 43–65. Oxford and New York: Oxford University Press.
Hammelstein, Philipp, and Steven Soifer. 2006. 'Is "Shy Bladder Syndrome" (Paruresis) Correctly Classified as a Social Phobia?' *Anxiety Disorders* 20: 296–311.
Hinshelwood, R.D. 1989. *A Dictionary of Kleinian Thought*. London: Free Association Books.
Hocquenghem, Guy. 1978. *Homosexual Desire*. Trans. Daniella Dangoor. London: Allison and Busby.
Horan, Julie L. 1996. *The Porcelain God: A Social History of the Toilet*. Toronto: A Birch Lane Press Book, Published by Carol Publishing Group.
Horkheimer, Max, and Theodor W. Adorno. 1972. *Dialectic of Enlightenment*. Trans. John Cumming. New York: Herder.
Houlbrook, Matt. 2000. 'The Private World of Public Urinals: London 1918–57.' *London Journal* 25.1: 52–70.
Hutcheon, Linda. 2000. *A Theory of Parody: The Teachings of Twentieth-century Art Forms*. New York: University of Illinois Press.
Ian, Marcia. 2001. 'The Primitive Subject of Female Bodybuilding: Transgression and Other Postmodern Myths.' *Differences: A Journal of Feminist Cultural Studies* 12.3: 69–100.

Ihde, Don. 2007. *Listening and Voice: Phenomenologies of Sound*. 2nd ed. Albany, NY: State University of New York Press.
Inglis, David. 2002. 'Dirt and Denigration: The Faecal Imagery and Rhetorics of Abuse.' *Postcolonial Studies* 5.2: 207–21.
Inglis, David, and Mary Holmes. 2000. 'Toiletry Time: Defecation, Temporal Strategies and the Dilemmas of Modernity.' *Time and Society* 9.2–3: 223–45.
Ingraham, Catherine. 1992. 'Initial Proprieties: Architecture and the Space of the Line.' In S*exuality and Space*, ed. Beatriz Colomina, 255–71. New York: Princeton University Press.
Irigaray, Luce. 1985. *This Sex Which Is Not One*. Trans. Catherine Porter. Ithaca, NY: Cornell University Press.
Isaksen, Lise Widding. 2002. 'Toward a Sociology of (Gendered) Disgust: Images of Bodily Decay and the Social Organization of Care Work.' *Journal of Family Issues* 23.7: 791–811.
Joyce, James. 1961[1922]. *Ulysses*. New York: Random House.
– 1976 [1916]. *A Portrait of the Artist as a Young Man*. New York: Penguin.
– 1976 [1939]. *Finnegans Wake*. New York: Penguin.
Juang, Richard M. 2006. 'Transgendering the Politics of Recognition.' In *The Transgender Studies Reader*, ed. Susan Stryker and Stephen Whittle, 706–19. New York: Routledge.
Kessler, Suzanne, J. 2002. *Lessons from the Intersexed*. New Brunswick, NJ, and London: Rutgers University Press.
Kipnis, Laura. 1996. *Bound and Gagged: Pornography and the Politics of Fantasy in America*. New York: Grove Press.
Kira, A. 1966. *The Bathroom: Criteria for Design*. New York: Center for Housing and Environmental Studies, Cornell University.
– 1976. *The Bathroom*. New York: Viking Press.
Kirby, K. 1996. *Indifferent Boundaries: Exploring the Space of the Subject*. New York: Guilford Press.
Kitchin, Rob, and Robin Law. 2001. 'The Socio-Spatial Construction of (In)accessible Public Toilets.' *Urban Studies* 38.2: 287–98.
Krafft–Ebing, Richard Von. 1965. *Psychopathia Sexualis: A Medico-Forensic Study*. Trans. Ernest van den Haag. New York: G.P. Putnam's Sons.
Kristeva, Julia. 1982. *Powers of Horror: An Essay on Abjection*. New York: Columbia University Press.
Kulley, Mike. 1970. *Johns in Europe: Toilet Training for Tourists*. Los Angeles: C/O Coraco.
Lacan, Jacques. 2006. 'The Instance of the Letter in the Unconscious, or Reason since Freud.' In *Écrites* by Jacques Lacan. Trans. Bruce Fink. New York and London: W.W. Norton and Company.

Lahiji, Nadir, and D.S. Friedman, eds. 1997. *Plumbing: Sounding Modern Architecture*. New York: Princeton Architectural Press.
Lambton, Lucinda. 2007. *Temples of Convenience and Chambers of Delight*. Stroud: Tempus Publishing Limited.
Lane, Christopher. 1999. 'Living Well Is the Best Revenge: Outing, Privacy, and Psychoanalysis.' In *Public Sex / Gay Space*, ed. William L. Leap. New York: Columbia University Press.
Laporte, Dominique. 2000. *History of Shit*. Trans. Nadia Benabid and Rodolphe el–Khoury. Cambridge, MA: MIT Press.
Largey, Gale Peter, and David Rodney Watson. 1972. 'The Sociology of Odors.' *The American Journal of Sociology* 77.6: 1021–34.
Leach, Neil. 1997. *Rethinking Architecture: A Reader in Cultural Theory*. New York: Routledge.
Lefebvre, Henri. 1998. *The Production of Space*. Trans. Donald Nicholson-Smith. Cambridge: Blackwell.
Lerley, Merritt. 1999. 'The Bathroom: An Epic.' *American Heritage* 50.3: 77–83.
Levinas, Emmanuel. 1989. 'Time and the Other.' In *The Levinas Reader*, ed. Sean Hand, trans. Richard A. Cohen, 37–58. Oxford: Blackwell.
Lewin, Ralph A. 1999. *Merde: Excursions in Scientific, Cultural, and Sociohistorical Coprology*. New York: Random House.
Lloyd, Genevieve. 1984. *The Man of Reason: 'Male' and 'Female' in Western Philosophy*. Minneapolis: University of Minnesota Press.
Logan, Peter Melville. 1997. *Nerves and Narratives: A Cultural History of Hysteria in 19th Century British Prose*. Berkeley and Los Angeles: University of California Press.
Longhurst, Robyn. 2001. *Bodies: Exploring Fluid Boundaries*. London and New York: Routledge.
Love, Brenda. 1992. *Encyclopedia of Unusual Sex Practices*. London: Greenwich Editions.
Lukes, H.N. 'Unrequited Love: Lesbian Transference and Revenge in Psychoanalysis.' In *Homosexuality and Psychoanalysis*, ed. Tim Dean and Christopher Lane, 250–65. Chicago and London: University of Chicago Press.
Magni, Sonia, and Vasu Reddy. 2007. 'Performative Queer Identities: Masculinities and Public Bathroom Usage.' *Sexualities* 10.2: 229–42.
Mallgrave, Harry Francis. 1997. In *Plumbing: Sounding Modern Architecture*, ed. Nadir Lahiji and D.S. Friedman, 123–36. New York: Princeton Architectural Press.
Marks, Laura U. 2000. *The Skin of the Film: Intercultural Cinema, Embodiment, and the Senses*. Durham and London: Duke University Press.
Masquelier, Adeline, ed. 2005. *Dirt, Undress, and Difference: Critical Perspectives on the Body's Surface*. Bloomington and Indianapolis: Indiana University Press.

Mattilda, aka Matt Bernstein Sycamore, ed. 2006. *Nobody Passes: Rejecting the Rules of Gender and Conformity*. Emeryville, CA: Seal Press.

Maugham, Somerset. 1930. *On a Chinese Screen*. New York: Doran.

Maynard, Steven. 1994. 'Through a Hole in the Lavatory Wall: Homosexual Subcultures, Police Surveillance, and the Dialectics of Discovery, Toronto, 1890–1930.' *Journal of the History of Sexuality* 5.21: 207–42.

– 1998. 'On the Case of the Case: The Emergence of the Homosexual as a Case History in Early-Twentieth-Century Ontario.' In *On the Case: Explorations in Social History*, ed. Franca Iacovetta and Wendy Mitchenson, 65–87. Toronto: University of Toronto Press.

McClintock, Anne. 1995. *Imperial Leather: Race, Gender and Sexuality in the Colonial Contest*. New York: Routledge.

McLaughlin, Terence. 1971. *Dirt: A Social History as Seen through the Uses and Abuses of Dirt*. New York: Stein and Day.

Meigs, Anna S. 1984. *Food, Sex, and Pollution: A New Guinea Religion*. New Brunswick, NJ: Rutgers University Press.

Menninghaus, Winfried. 2003. *Disgust: Theory and History of a Strong Sensation*. Trans. Howard Eiland and Joel Golb. New York: State University of New York Press.

Merleau-Ponty, Maurice. 1968. *The Visible and the Invisible*. Trans. Alphonso Lingis. Evanston, IL: Northwestern University Press.

Miller, Susan B. 2004. *Disgust: The Gatekeeper Emotion*. London: Analytic Press.

Miller, William Ian. 1997. *The Anatomy of Disgust*. Cambridge, MA: Harvard University Press.

Mitchell, David T., and Sharon L. Snyder. 2000. *Narrative Prosthesis: Disability and the Dependencies of Discourse*. Ann Arbor: University of Michigan Press.

Molesworth, Helen. 1997. 'Bathrooms and Kitchens: Cleaning House with Duchamp.' In *Plumbing: Sounding Modern Architecture*, ed. Nadir Lahiji and D.S. Friedman, 75–92. New York: Princeton Architectural Press.

Moore, Alison. 2004. '*Kakao* and *Kaka*: Chocolate and the Excretory Imagination of Nineteenth-Century Europe.' In *Cultures of the Abdomen: Diet, Digestion, and Fat in the Modern World*, ed. Christopher E. Forth and Ana Carden-Coyne, 51–69. New York: Palgrave Macmillan.

– 2008. 'Fin de Siècle Sexuality Excretion.' In *Fin de Siècle Sexuality: The Making of a Central Problem*, ed. Peter Cryle and Christopher Forth, 125–39. Newark, DE: University of Delaware Press.

– 2009. 'Colonial Visions of "Third World" Toilets: A Nineteenth-century Discourse That Haunts Contemporary Tourism.' In *Ladies and Gents: Public Toilets and Gender*, ed. Olga Gershenson and Barbara Penner, 97–113. Philadelphia: Temple University Press.

Morgan, Margaret. 2002. 'The Plumbing of Modern Life.' *Postcolonial Studies* 5.2: 171–95.
Morin, Jack. 1998. *Anal Pleasure and Health*. San Francisco: Down There Press.
Morrison, Toni. 1992. *Playing in the Dark: Whiteness and the Literary Imagination*. New York: Vintage.
Morna, E. Gregory, and Sian James. 2006. *Toilets of the World*. London and New York: Merrell.
Muñoz, José Esteban. 1999. *Disidentifications: Queers of Color and the Performance of Politics*. Minneapolis and London: University of Minnesota Press.
– 2007. 'Cruising the Toilet: LeRoi Jones / Amiri Baraka, Radical Black Traditions and Queer Futurity.' *GLQ* 13.2–3: 353–67.
Namaste, Viviane K. 2000. *Invisible Lives: The Erasure of Transsexual and Transgendered People*. Chicago: University of Chicago Press.
Nast, Heidi J. 1998. 'Unsexy Geographies.' *Gender, Place, and Culture* 5: 191–206.
– 2000. 'Mapping the "Unconscious": Racism and the Oedipal Family.' *Annals of the Association of American Geographers* 90.2: 215–55.
Nataf, Zachary I. 2006. 'Lesbians Talk Transgender.' In *The Transgender Studies Reader*, ed. Susan Stryker and Stephen Whittle, 439–48. New York: Routledge.
Noble, Jean Bobby. 2004. *Masculinities without Men? Female Masculinity in Twentieth-Century Fictions*. Vancouver: University of British Columbia Press.
– 2006. *Sons of the Movement: FTMs Risking Incoherence on a Post-Queer Cultural Landscape*. Toronto: Women's Press.
Obeng, Samuel Gyasi. 2000. 'Speaking the Unspeakable: Discursive Strategies to Express Language Attitudes in Legon (Ghana) Graffiti.' *Research on Language and Social Interaction* 33.3: 291–319.
Oliver, Kelly. 2004. *The Colonization of Psychic Space: A Psychoanalytic Social Theory of Oppression*. Minneapolis: University of Minnesota Press.
Oliver, Kelly, and Benigno Trigo. 2003. *Noir Anxiety*. Minneapolis: University of Minnesota Press.
Penner, Barbara. 2005. 'Researching Female Public Toilets: Gendered Spaces, Disciplinary Limits.' *Journal of International Women's Studies* 6.2: 81–98.
– 2009. '(Re)Designing the "Unmentionable": Female Toilets in the Twentieth Century.' In *Ladies and Gents*, ed. Olga Gershenson and Barbara Penner, 141–50. Philadelphia: Temple University Press.
Phillips, Adam. 2006. *Side Effects*. New York: Harper Perennial.
Philo, Chris, and Hester Parr. 2003. 'Introducing Psychoanalytic Geographies.' *Social and Cultural Geography* 4.3: 283–93.
Pike, David L. 2005. 'Sewage Treatments: Vertical Space and Waste in Nineteenth-Century Paris and London.' In *Filth: Dirt, Disgust, and Moral Life*, ed.

William A. Cohen and Ryan Johnson, 51–77. Minneapolis and London: University of Minnesota Press.
Pile, Steve. 1996. *The Body and the City: Psychoanalysis, Space and Subjectivity.* New York: Routledge.
Pilling, Meredith. 2006. *Queer Encounters: Exploring Experiences of (Gender) Queers in Women's Public Washroom Spaces.* Unpublished MA thesis, Brock University, St Catharines, ON.
Pronger, Brian. 1999. 'Outta My Endzone: Sport and the Territorial Anus.' *Journal of Sport and Social Issues* 23.4: 373–89.
Prosser, Jay. 1998a. *Second Skins: The Body Narratives of Transsexuality* New York: Columbia University Press.
– 1998b. 'Transsexuals and the Transsexologists: Inversion and the Emergence of Transsexual Subjectivity.' In *Sexology in Culture: Labelling Bodies and Desires*, ed. Lucy Bland and Laura Doan, 116–31. Chicago: University of Chicago Press.
Rawes, Peg. 2007. *Irigaray for Architecture.* Thinkers for Architects Series. London and New York: Routledge, Taylor, and Francis Group.
Reyburn, Wallace. 1969. *Flushed with Pride: The Story of Thomas Crapper.* London: Pavilion Books Limited.
Reynoldson, Fiona. 1994. *Victorian Bathrooms.* Harlow, Essex: Longman Group.
Rich, Adrienne. 1980. 'Compulsory Heterosexuality and Lesbian Existence.' In her *Blood, Bread, and Poetry: Selected Prose, 1979–1985*, 23–75. New York: W.W. Norton and Company.
Rockefeller, Abby A. 1998. 'Civilization and Sludge: Notes on the History of the Management of Human Excreta.' *Capitalism, Nature, Socialism* 9.3: 3–18.
Rose, Gillian. 1993. *Feminism and Geography: The Limits of Geographical Knowledge.* Cambridge: Polity.
– 1996. 'As If the Mirrors Had Bled: Masculine Dwelling, Masculinist Theory and Feminist Masquerade.' In *BodySpace: Destabilizing Geographies of Gender and Sexuality*, ed. N. Duncan, 56–74. London: Routledge.
Rosolato, Guy. 1974. 'La voix: Entre corps et langage.' *Revue Francaise de Psychanalyse.* 37.1: 81.
Rubin, Gayle. 2006. 'Of Catamites and Kings: Reflections on Butch, Gender and Boundaries.' In *The Transgender Studies Reader*, ed. Susan Stryker and Stephen Whittle, 471–81. New York and London: Routledge.
Rush, Fred. 2009. *On Architecture: Thinking in Action.* New York: Routledge, Taylor and Francis Group.
Sacco, Lynn. 2002. 'Sanitized for Your Protection: Medical Discourse and the Denial of Incest in the United States, 1890–1940.' *Journal of Women's History* 14.3: 80–104.

Sade, Marquis de. 1966. *The 120 Days of Sodom and Other Writings*. Trans. A. Michelson. New York: Grove Press.
– 1968. *Juliette*. Trans. A. Wainhouse. New York: Grove Press.
Salamon, Gayle. 2004. 'The Bodily Ego and the Contested Domain of the Material.' *differences: A Journal of Feminist Cultural Studies* 15.3: 95–122.
Salzberger-Wittenberg, Isca. 1970. *Psycho-Analytic Insight and Relationships: A Kleinian Approach*. New York: Routledge and Kegan Paul.
Schilder, Paul. 1950. *The Image and Appearance of the Human Body: Studies in the Constructive Energies of the Psyche*. New York: International Universities Press.
Sedgwick, Eve Kosofsky. 1990. *Epistemology of the Closet*. Berkeley and Los Angeles: University of California Press.
– 2003. *Touching Feeling: Affect, Pedagogy, Performativity*. Durham and London: Duke University Press.
Serano, Julia. 2007. *Whipping Girl: A Transsexual Woman on Sexism and the Scapegoating of Femininity*. Emeryville, CA: Seal Press.
Shildrick, Margrit. 1997. *Leaky Bodies and Boundaries: Feminism, Postmodernism and (Bio)ethics*. New York: Routledge.
– 'Prosthetic Performativity: Deleuzian Connections and Queer Corporealities.' In *Deleuze and Queer Theory*, ed. Chrysanthi Nigiani and Merl Storr, 115–33. Edinburgh: Edinburgh University Press.
Sibley, David. 1992. 'Outsiders in Society and Space.' In *Inventing Places: Studies in Cultural Geography*, ed. K. Anderson and F. Gale, 107–22. London: Longman.
– 1995. *Geographies of Exclusion: Societies and Difference in the West*. London: Routledge.
– 2003. 'Geography and Psychoanalysis: Tensions and Possibilities.' *Social and Cultural Geography* 4.3: 391–9.
Silverman, Kaja. 1988. *The Acoustic Mirror: The Female Voice in Psychoanalysis and Cinema*. Bloomington and Indianapolis: Indiana University Press.
– 1996. *The Threshold of the Visible World*. New York and London: Routledge.
Sivulka, Juliann. 2001. 'From Domestic to Municipal Housekeeper: The Influence of the Sanitary Reform Movement on Changing Women's Roles in America, 1860–1920.' *Journal of American Culture* 22.4: 1–7.
Soyinka, Wole. 1970 [1965]. *The Interpreters*. Portsmouth: Heinemann.
Srinivas, Tulasi. 2002. 'Flush with Success.' *Space and Culture* 5.4: 368–86.
Stallybrass, Peter, and Allon White. 1986. *The Politics and Poetics of Transgression*. Ithaca, NY: Cornell University Press.
– 2007. 'The City: The Sewer, the Gaze, and the Contaminating Touch.' In *Beyond the Body Proper: Reading the Anthropology of Material Life*, ed. Margaret

Lock and Judith Farquhar, 266–85. Durham and London: Duke University Press.

Stoler, Ann Laura. 1995. *Race and the Education of Desire: Foucault's History of Sexuality and the Colonial Order of Things*. London: Duke University Press.

Stone, Sandy. 2006. 'The Empire Strikes Back: A Posttranssexual Manifesto.' In *The Transgender Studies Reader*, ed. Susan Stryker and Stephen Whittle, 221–35. New York and London: Routledge.

Stryker, Susan. 2006. 'My Words to Victor Frankenstein Above: The Village of Chamounix: Performative Transgender Rage.' In *The Transgender Studies Reader*, ed. Susan Stryker and Stephen Whittle, 244–56. New York: Routledge.

– 2008. *Transgender History*. Berkeley, CA: Seal Press.

Stryker, Susan, and Stephen Whittle, eds. 2006. *The Transgender Studies Reader*. New York and London: Routledge.

Sullivan, Nikki. 2003. *A Critical Introduction to Queer Theory*. New York: New York University Press.

Tewksbury, Richard. 2004. 'The Intellectual Legacy of Laud Humphreys: His Impact on Research and Thinking about Men's Public Sexual Encounters.' *The International Journal of Sociology and Social Policy* 24.3–5: 32–57.

Thomas, Calvin. 1996. *Male Matters: Masculinity, Anxiety, and the Male Body on the Line*. Urbana: University of Illinois Press.

– 2008. *Masculinity, Psychoanalysis, Straight Queer Theory: Essays on Abjection in Literature, Mass Culture, and Film*. New York: Palgrave Macmillan.

Tomes, Nancy. 2001. 'The History of Shit: An Essay Review.' *Journal of the History of Medicine* 56: 400–1.

Tremain, Shelley, ed. 2005. *Foucault and the Government of Disability*. Ann Arbor: The University of Michigan Press.

Trotter, David. 2005. 'The New Historicism and the Psychopathology of Modern Life.' In *Filth: Dirt, Disgust and Modern Life*, ed. William A. Cohen and Ryan Johnson, 30–48. London and Minneapolis: University of Minnesota Press.

Valentine, David. 2007. *Imagining Transgender: An Ethnography of a Category*. Durham and London: Duke University Press.

Valverde, Mariana. 1991. *The Age of Light, Soap and Water: Moral Reform in English Canada, 1885–1925*. Toronto: McClelland and Stewart.

Van Der Geest, Sjaak. 2002. 'The Night-Soil Collector: Bucket Latrines in Ghana.' *Postcolonial Studies* 5.2: 197–206.

Vasseleu, Cathryn. 1998. *Textures of Light: Vision and Touch in Irigaray, Levinas and Merleau-Ponty*. New York: Routledge.

Vigarello, Georges. 1988. *Concepts of Cleanliness: Changing Attitudes in France since the Middle Ages*. Trans. Jean Birrell. New York: Cambridge University Press.

Wachholz, Sandra. 2005. 'Hate Crimes against the Homeless: Warning-Out New England Style.' *Journal of Sociology and Social Welfare* 32.4: 141–63.
Wallon, Henri. 1934. *Les origines du caractère chez l'enfant: Les préludes du sentiment de personnalité*. Paris: Boivin et Cie.
Warner, Michael. 2002. *Publics and Counterpublics*. New York: Zone Books.
Waugh, Thomas. 2007. *The Romance of Transgression in Canada*. Montreal and Kingston: McGill-Queen's University Press.
Wedd, Kit. 1996. *The Victorian Bathroom Catalogue: A Treasury of over 1,000 Baths, Showers, Fixtures and Fittings*. London: Studio Editions.
Weinberg, Martin S., and Colin J. Williams. 2005. 'Fecal Matters: Habitus, Embodiments, and Deviance.' *Social Problems* 52.3: 315–36.
Whittle, Stephen. 2006. 'Foreword.' In *The Transgender Studies Reader*, ed. Susan Stryker and Stephen Whittle, xi–xvi. New York: Routledge.
Williams, Linda. 1989. *Hard Core: Power, Pleasure, and the 'Frenzy of the Visible.'* Berkeley: University of California Press.
Williams, M.T. 1991. *Washing 'The Great Unwashed': Public Baths in Urban America 1840–1920*. Columbus: Ohio State University Press.
Wills, David. 1995. 'Rome, 1985.' In his *Prosthesis*. Stanford, CA: Stanford University Press.
Wilton, R.D. 1998. 'The Constitution of Difference: Space and Psyche in Landscapes of Exclusion.' *Geoform* 29.2: 173–85.
Winkler, Gail Caskey, and Roger W. Moss. 1984. 'How the Bathroom Got White Tiles ... and Other Victorian Tales.' *Historic Preservation* 36.1: 32–5.
Wright, Lawrence. 1960. *Clean and Decent: The Fascinating History of the Bathroom and the WC*. Toronto: University of Toronto Press.
Young, Iris. 1990. 'The Scaling of Bodies and the Politics of Identity.' In her *Justice and the Politics of Difference*, 122–55. Princeton, NJ: Princeton University Press.
Young-Bruehl, Elisabeth. 1996. *The Anatomy of Prejudices*. Cambridge, MA: Harvard University Press.
Zimring, Carl. 2004. 'Dirty Work: How Hygiene and Xenophobia Marginalized the American Waste Trades, 1870–1930.' *Environmental History* 9.1. 80–101.
Žižek, Slavoj. 1997. *The Plague of Fantasies*. London: Verso.

Index

abject feminine body, 50–1, 81–2, 108–9, 127–8, 244nn12–13
abjection: about gender coherence, 135–6; of bodily fluids, 136–51, 146–7; and cultural infection, 142–3; and defecation, 166; and eroticization of the toilet, 204; and gendering of bodily ego, 208; modernity's production of, 141; 'scatontological anxiety,' 50–1; the term, 137
absorb and absorption, 148–9
acoustic mirror, 106–9, 110–13, 117–18, 125–8, 241–2n1, 243n8
Acoustic Mirror, The, 125. *See also* Silverman, Kaja
acoustic sensations: class and, 11–12; and codes of silence, 185; elimination and sex linked through, 7; feelings associated with, 112; and feminine (maternal) speech, 117–18, 124; historic context, 243n9; of peeing (urinary echoes), 125–6; and performative non-speech, 121; performative unhearing, 116–17; policing of, 6; seeing gender with our ears, 110–11; and signifiers of sexual difference, 114; and surveillance, 105–6; voice and urinary acoustics, 111, 114–21, 123
activist groups for gender rights. *See* law and legislation
Adorno, Theodor W., 167
advertisements in toilets, 148, 179, 254nn11–12
Advocates for Children (New York), 230n5
anal eroticism and elimination: defecation and penetration, 161–2, 164–6, 247nn11–12; and Freud, 31, 249–50nn23–24; and normative body politics, 172; and uprightness, 207–8. *See also* desire; eroticism and elimination
Anderson, Richard, 236n17
Anspaugh, Kelly, 248n14
anus: as abject, 141; (be)hindsight, 183; metonymic relationships, 252n4; and urinary urinal positions, 183–5
anxiety: about human mortality, 40, 42; about sodomy, 10; cultural, political context of, 3–4; heterosexual genitalia, 175–6; public washroom

as site of, 5; 'scatontological anxiety,' 50–1; transphobia and homophobia, 162
Anzieu, Didier, 48, 98, 242n4
architecture and design (of washrooms): abject feminine body in, 81–2, 108–9; to absorb loss and impotence, 108, 206–8; acoustic design, 106–7, 110–12, 117; and 'anatomy of detail,' 79; and bedrooms, 30–1; and colonialism, 246n9; and disability, 253n10; as disciplining gender, 5–6; and doors on public toilets, 177–8, 253n6, 253n9; and enacting disgust, 152; and gendering of stink, 153–6; history of public toilets and sex in, 28, 169–70 (*see also* history and cultural politics of excretion); homophobic spatial configurations, 180; of the ideal bathroom, 216–17; identity confusion through, 118–19; and illusion of neutrality, 32, 52; incorporating alternative images, 46–7; in institutionalization of heterosexuality, 8; intolerance for the feminine in, 82; as masculine/feminine, 128; and plumbing, 209; prioritizing sensory systems in, 136, 141, 151; racialization and class in, 11–13, 61, 251n30; reworking as queer and trans-positive, 212–13; use of colour in, 139–40 (*see also* white and whiteness); vertical and horizontal meanings, 81–2, 208. *See also* choreography of the body; panopticism; urinals, male; urinals and urinettes, female
Aretxaga, Begona, 247–8n14

automated sanitary technology, 167–8

Bailey, Peter, 106
Bally Total Fitness Club (Worcester, Massachusetts), 231n7
Bataille, Georges, 40, 199
bathtub, 142, 145
Beach Place Ventures Limited, 9
bedrooms and toilets, 30–1
Bentham, Jeremy, 82, 84, 105, 233n6
Benthien, Claudia, 62, 258n4
Berlant, Lauren, 173, 204
Bersani, Leo, 202, 252n4
bidet, 38, 251n30
binary gender axis. *See* sexual difference
bio-political regulation of the body, 10, 24, 28, 30, 86, 171, 212. *See also* science
blood, the symbolic of, 150. *See also* menstruation
bodily ego and gender, 43–51; and bathroom encounters, 48–9; and the bathroom mirror, 96–103; and constancy of bathroom fixtures, 144; relationship with the abject, 208; role of orifices in, 47–8; the term, 45; territorial occupying of space, 121–3. *See also* boundaries (body); psychoanalysis
'border intimacies,' 204
boundaries (body): and constancy of bathroom fixtures, 144; and disgust with urine, 143; heterosexual angst of, 31–2; masculine/feminine differences, 246–7n11; and sex in public toilets, 203; skin ego, 48, 242n4; territorialization and defecation, 163; unstable in face of shit, 29

bourdalous, 37
Bourke, John G. (*Scatalogic Rites of All Nations*), 31, 33–4, 244n14, 247n12, 250–1n25
Boyd, Nan Alamilla, 66
Braham, William, 142
breasts, 66–7, 71, 98
British Columbia Human Rights, 9, 230n5
Britzman, Deborah, 210
Bronfen, Elisabeth, 213
Butler, Judith, 7–8; abjection, 50–1, 137; and bodily ego, 45; and Foucault, 43, 96–7; *Giving an Account of Oneself*, 121; heterosexual matrix, 171–2; loss and prohibition, 40, 41, 195, 213; mistaken gender, 53; nonviolent responses, 68; subjection, 87; unconscious identification, 63, 89; *Undoing Gender*, 237n3; upsetting the norm, 66; violent responses, 78

Cahill, Spencer, 33
California Student Safety and Violence Prevention Act (2000), 230n5
Camfield, William, 240n1
capitalism: alienation and sanitary technology, 167; and design of washrooms, 139–40; in history of public toilets, 28, 33, 234n8; and queer sex in toilets, 173
castration: creating space of, 128; and display of the penis, 121; feminine speech and, 117–19; and racialized other, 183; and sound, 112; and transgender, transsexual, and/or genderqueer, 109; use of term, 242n3
censorship, 227n1
cesspools, 40–1, 236n18

Chadwick, Edwin, 79
children: and anal eroticism, 250nn23–4; fear of abduction and paedophilia, 41, 57–8, 190–3; and gender signs on doors, 233n5; and homophobic panic, 190–3; and incest, 176; LGBTI people read as threats to, 57; and mother's voice, 241–2n1. *See also* toilet training
chocolate, 36, 251n25
choreography of the body, 21; and anatomization of sex, 145–6; and anus at the urinal, 183–5; and authorized orifices and fluids, 172; class and genital organization, 80; of feet in stalls, 93–5; investment in the norm and, 97; managing desire and disgust, 128; stink as masculine, 153; and the urinal, 80–1, 122, 129, 219–20; urinary positions, 81, 128–33. *See also* urinals, male; urinals and urinettes, female
Christianity: cleanliness next to godliness, 34, 144–5; and hygienic superego, 145; and menstrual blood, 147; on queer sex in toilets, 173
Chun, Allen, 251n30
cissexuals: cissexist privilege, 53–4, 58–9; use of term, 8
Clark, John R., 248n14
class: in design, 13, 233n4, 248n16; domination through faecal comparisons, 249–50n23; elimination by Victorian women and, 39; and history of ideas of cleanliness, 234n11; policing of gender and hygiene, 6–7, 79; in toilet technologies, 234n9, 236n20; white

bourgeois subject formation, 42; working-class male bodies, 236n17
cleanliness: of bathrooms and gender, 137–8; clean and unclean as co-dependent, 145; and colour, 86, 139; and employment and housing, 245n3; history of ideas of, 234n11; next to godliness, 34, 144–5. *See also* hygiene and hygienists
clitoris. *See* vagina and clitoris
close stools, 37, 243n9
Cockayne, Emily, 106
colonialism, 11, 61, 233–4n7; and colour of toilets, 139; disciplinary institutions and, 86; domination through faecal comparisons, 249–50n23; ethic of gender purity, 140; and production of the abject, 141; role of excretion in history of, 33; and urinary positions, 244n14; in washroom design, 246n9
commodes, 37
communal activities, 34–5
Connor, Steven, 109–10
Cooper, Annabel, 33, 39
Cooper, Patricia, 31–2
coprophagia, 33, 36, 251n25
coprophilia, 141, 172, 251n25
Corbin, Alain, 245n2, 249n22
Crapper, Thomas, 243n9

Davidson, Michael, 102
da Vinci, Leonardo, 35
death: associated with excretion, 40–1, 140, 166–7; and menstrual blood, 147, 151; and whiteness, 246n5
de-idealizing gaze, 87–8
delayed elimination, 38–9, 132–3, 162

design of toilets. *See* architecture and design (of washrooms)
desire: and abjection, 136; and graffiti, 187; managing desire and disgust, 128, 160, 244n13, 247n12; and prohibition, 195–6, 199–200; public washroom as site of, 5; and unauthorized orifices, 172; use of the grotesque in, 204. *See also* anal eroticism and elimination
disability, 49–50, 101–3, 102, 251n31, 253n10
disease and dirt: associated with social body, 32–3; associated with street excretion, 39–40, 134; and class, 248n16; and cultural infection, 142; and gender purity, 7; and gender-segregated toilets, 63, 134–5; gender variance perceived as contagious, 64, 135; HIV and AIDs, 135, 245n1; quarantine effect, 93
disidentifications: homophobic and transphobic, 10; policing gender through, 137; pretence of fear masking, 76; as projective, 13, 51, 141; the term, 49. *See also* identity and identity categories
dreams of excrement, 36
Duchamp, Marcel *(Fountain)*, 82, 240n1
Duggan, Lisa, 13
Dyer, Richard, 139, 246n5, 255n15

Eaton, Timothy, 28
Edelman, Lee: (be)hindsight, 183–4; on homophobic fear, 10, 164, 193; reproductive futurity, 57, 172, 176; sinthomosexual, 177, 257n22; on unproductive sex, 252n2

Edwards, Donald, 246n4
ego. *See* bodily ego and gender
ejaculate, 163, 251n27
elimination and excretion. *See* anal eroticism and elimination; delayed elimination; eroticism and elimination; history and cultural politics of excretion
Elliot, Patricia, 44
Ellis, Havelock, 38–9
Eng, Doug, 182–3
Engels, Frederick, 249n22
eroticism and elimination: as abject desires, 136; cataloguing of sexual practices, 31; Freud and, 35–6, 235n14; historic examples of, 235n15, 250–1n25; the link of, 7; as matters of life and death, 42; and presence of faeces, 165–6; and sex in bathrooms, 168; and taboo, 199–200; and women's orifices, 35. *See also* anal eroticism and elimination

family (heteronormative): family toilets not exclusive to, 217; homosexual and trans subject as threat to, 176–7; and queer sex in toilets, 173; threatened by sexuality nonconformists, 190–1. *See also* reproduction
Fanon, Frantz (*Black Skin, White Masks*), 49
Feleski, Rita, 3
feminist theory, 44
fetishism, 30, 36
film, 102, 107–8, 139; masculine/feminine in Hollywood, 111–12, 113, 116, 124, 242n7
Foucault, Michel, 5; anxiety about gender variance, 4; 'art of light,' 85; *Discipline and Punish*, 96; 'great confinement,' 134; *History of Sexuality*, 27–43, 32, 43, 86; homosexuality, 169–70, 202; panopticism, 5–6, 80, 92–3; 'pedagogy of examination,' 42–3; role of psychoanalysis, 43; stall design, 253n6; surveillance, 55, 79; the symbolic of blood, 150; technologies of sex, 171
Frankfurt School, 167
Frascari, Marco, 3, 258n3
Freud, Sigmund: anal eroticism, 31, 249–50nn23–24; bodily ego, 45, 48, 97; *Civilization and Its Discontents*, 35–6, 207; cleanliness, 249–50n23; defecation, 164; human uprightness, 207; identification, 83, 210; racialization of the vagina, 125; sour grapes, 244n13; the unconscious, 29–30; uniting the sexual and the excremental, 235n14. *See also* Oedipal complex
Friedman, D.S., 6, 82, 135, 209, 210
Fuss, Diana, 83, 84, 210

gay bars, 115, 202
gay, lesbian, and bisexual (not trans) people: critical and intolerant of trans people, 13 (*see also* transphobia and homophobia); on defecating in public washrooms, 165; identity categories, 14; urinary positions of, 131; use of terminology, 19–20. *See also under* interviewees
gender identity anxiety. *See* anxiety
Gender Neutral Bathroom Survey (Transgender Law Center), 227n2
gender-neutral lavatories: about validating gender identities,

211–12; arguments for and against, 74, 211–13; and the ideal bathroom, 215–18; implemented at universities, 228n2; for people with physical disabilities, 237n2; at schools, 230n5
gender panic, 76–8
genderqueer people, 16. *See also* transgender, transsexual, and/or genderqueer
gender recognition (achieving): and apology for misreadings, 68; in bathrooms, 57, 59; and identifications, 210–11; integral to participation, 62; and mirrorical returns, 66. *See also* misrecognitions
gender-segregated toilets: choosing, 63, 239n9; dominant safety narratives justifying, 73; fear of 'men's' room, 22–3; heterosexism informing, 175; history of, 7, 28; laws governing, 70; policing by security guards, 72; sex-segregated toilets, 21; use of term, 21. *See also* public washrooms
gender signs on bathroom doors: and accessibility, 214–15; and children, 233n5; in defence against disease, 40; disciplining orifices, 30; and fear of abjection, 51; on the ideal bathroom, 215; and pedagogy of examination, 43; redesigning of, 212, 214–15; in reversibility in visual exchange, 65; 'urinary segregation,' 32. *See also* public washrooms
gender trickery, 69, 90
genitals: acoustic washroom design and, 110; governed by superegos, 136; inferred by sound, 111–12; and people with physical disabilities, 102; sewer systems compared to sex organs, 37–8, 42, 236n16; touching of in bathrooms, 31–2; vaginal mirage, 125–8; as visible/invisible, 112–13; voices and female, 119. *See also* penis; vagina and clitoris
Goffman, Irving, 33
Gosine, Andil, 255n14
graffiti, 186–90, 221, 256n18
Grand Central Station (New York), 175, 228–9n3. *See also* New York City
'Great Unwashed,' 43
Greed, Clara, 253n7

Halberstam, Judith, 52, 202
Hale, C. Jacob, 150, 249n20
hand washing, 138, 246n4
haptic visuality, 159–61
Harington, John *(The Metamorphosis of Ajax)*, 227n1
hearing and noise. *See* acoustic mirror; acoustic sensations
heterosexual matrices of the toilet, 174–86
heterosexual matrix (Butler), 7–8, 171, 176
Hispanic AIDS Forum (New York), 8
history and cultural politics of excretion, 3–4; academia's opinion of, 27; cesspool deaths, 40–1; covering sounds, 243n9; eroticism in, 247n12, 250–1n25; female urinals in, 220; in literature, 33–4; and masculine/feminine smells, 154; not always private, 201; and olfactory tastes, 249n22; panopticism in, 5–6; of Paris and London, 28–9; sex and elimination linked in, 35;

street excretion, 38–40; surveillance and policing, 79–80; treatment of excrement in, 247–8n14; and urinary positions, 244n14. *See also* sewer systems; Victorian era; water closet

HIV and AIDs, 135, 245n1. *See also* disease and dirt

Hollywood. *See under* film

Holmes, Mary, 245n15

home and heteronormativity, 44–5; and homelessness, 245n3; and the ideal bathroom, 217, 258n3, 258n4

Horan, Julie L., 235n15, 243n9

Horkheimer, Max, 167

Houlbrook, Matt, 185

Hugo, Victor (*Les Misérables*), 29

humane recognitions, 53, 58, 62, 68, 103, 211, 221

Hutcheon, Linda, 104

hygiene and hygienists: and cultural infection, 142; efficiency and economy of, 233n6; excrement and loss, 209; as gendered, 138; gender purity and, 135–6; hygiene rituals and touch, 157; and modernity, 32; policing hygiene, 7, 42, 79; relation to colour, 139; and visual integrity, 142. *See also* cleanliness; sanitary engineers and city planners

hygienic superego, 135–6, 144–5, 206–7

hygrophilia, 136

ideal bathroom, 215–18, 258n3

identity and identity categories: bodily ego and, 44–51, 97, 98; cissexual confusion through trans encounters, 65–6, 68–9, 238n8; confusion through lavatory design, 118–19; as defence against subject dissolution, 40; difference between gender and sex, 14; explanation of, 13–14; gender misreadings, 53–4, 56–7, 59–60; gender performances in washrooms, 67, 152; gender pronouns and, 249n20; and gender stasis, 109; identifications at a distance (Silverman), 68; internal self-portrait and, 98; managing of, 210–11; mirrorical return, 83; and public participation, 237n3; response to loss, 206, 210; and the return of the abject, 209–10; role of orifices in, 47–8; sexual and gender identifications, 14; shame in misreadings, 60–2; and trauma, 213; unstable in face of shit, 29. *See also* disidentifications

Ihde, Don, 116

incest, 176

infantalization, 231n7

Inglis, David, 137, 231–2n10, 245n15, 249n23

Ingraham, Catherine, 32

interviews (methodology), 5–6, 13, 20–4, 21–2; confidentiality of identity, 13, 232n20; use of identity categories in, 14, 19–20

interviewees: on anxiety of voice and urinary acoustics, 111, 123; on association between excretion and sex, 174–86; on bathroom cleanliness, 137–8; on custodial work aligned with dirt, 245n3; on delayed elimination, 132–3; on feet in stalls, 93–5; on gender-neutral toilets, 211–12; on the ideal bathroom, 215–18, 257–8n2; on imaginary penis for 'men's' room, 122; on looks in 'men's' toilets, 90; on menstruation, 126–7, 147–50; on mirrors, 86–7; on peeing, 122–3; on purification rituals, 157–9; on racist imagery and acoustics, 120–1; on recognition as trans (not male or female), 59; on smells, 152–3; on sound in toilets, 114–17; on speech in toilets, 114; status as castrated subjects, 109; on urinals, 219; on urinary positions, 129, 131; on use of stalls, 93, 240n3; on voice for access to toilets, 119–20

- anonymous or not named: on gender-neutral toilets, 211–12; on mirror avoidance, 100; on perceived threat to children, 191; on predators, 75; on safety, 73; on transphobia, 76; on washroom looks, 89–90
- bisexual: on bathroom cleanliness, 137–8; on colour of toilets, 139
- female-to-male trans (NYC): on being arrested, 71
- femme: on misreadings as predators, 75
- gay male: on racism of gay male public sex culture, 182
- genderqueer: on reversibility in visual exchange, 65
- non-trans gay male: on perceived threat to children, 192, 193; on policing sex in toilets, 177
- non-trans queer femme: on gender-neutral toilets, 211; on heterosexual genitalia anxiety, 175–6
- queer: on the ideal bathroom, 257n2
- queer male: on femininity and smells, 153
- trans: expressions of hate as gendered, 77; on heterosexual construction of masculinity, 74; on heterosexual genitalia anxiety, 175; on homophobia and washrooms, 169
- trans masculine: on menstrual products, 149–50; on urinary positions, 129
- trans men: on bathroom cleanliness, 137–8; on gender distinctions, 69; on heterosexual construction of masculinity, 74; impact of using gender-segregated toilets, 60; on masculine/feminine smells, 154, 155; on mirrors in 'men's' toilets, 90; on misreadings as predators, 75; on remains in the toilet, 143; on urinary positions, 129; on using 'men's' room, 96
- trans, two-spirited (Vancouver): on being arrested, 71; on delayed elimination, 132; on the ideal bathroom, 217; impact of using gender-segregated toilets, 60–1
- trans woman: on being arrested, 72
- Brant (Scottish and French, non-trans queer man; Toronto): on choosing a urinal, 180; on code of silence, 185; on homophobic fear of exposure, 185; on public sex, 253n9; on racism of gay male public sex culture, 182, 254–5n13; on sex in public, 177–8, 198, 201
- Bryan (Irish, white, transmasculine, queer; San Francisco): on gender policing by non-trans men,

72, 239n10; on heteronormativity of bathrooms, 176; on urinals, 219
- Butch Coriander (WASP, non-trans woman, genderqueer, butch-dyke; Toronto): on apologies by cissexuals, 68–9; on colour of toilets, 139; on gender policing, 56
- Callum (white, trans man, queer; Toronto): on ads in toilets, 179, 254n11; on defecating in public washrooms, 162; on leaving stalls, 95; on meaning of 'clean,' 159; on menstrual products, 150; on misreadings of predators, 75; on security guards, 70; on urinary positions, 131; on using 'men's' room, 96; on voice for access to toilets, 119; on washroom line-ups, 103–4; on washroom mirrors, 88, 90
- Carol: on menstrual products, 149
- Carol Queen (northern-European, non-trans female, bisexual, polyamorous; San Francisco): on graffiti, 187; on sex in public, 200
- Charlie (Jewish, non-trans genderqueer male, bisexual, polyamorous; San Francisco): on graffiti, 186, 189
- Chloe (white, non-trans queer femme; Vancouver): on cleanliness, 245n1, 245n3; on construction of masculinity, 74; on defecating in public washrooms, 162; on erotic looks, 89; on femininity and smells, 153; on menstrual products, 149; on purification rituals, 158; on urinary positions, 130, 131; on washroom mirrors, 88
- Claude (British and French, non-trans, butch, lesbian; London, ON): on eroticizing toilets, 200; on gender recognition in bathrooms, 57
- Cole (two-spirited, trans): on urinals, 219
- Crystal (Chinese-Canadian, non-trans genderqueer woman, lesbian; Toronto): on the ideal bathroom, 216; on public gay sex culture, 181; on urinals, 219
- David (European, Christian, non-trans man, gay; Kingston, ON): on character of toilets, 159; on fear-based protocol, 180; on gender categorizing, 55; on gender-neutral toilets, 211; on heteronormativity of bathrooms, 176; on homophobic fear of exposure, 186; on peeing, 123; on pornography in toilets, 179; on speech in toilets, 114
- Diane (white, non-trans woman, androgynous, lesbian): on choosing gender-segregated toilets, 239n9; on design of toilets, 84–5; on graffiti, 188; on lack of odours, 152
- Dorothy (white, genderqueer): on choosing gender-segregated toilets, 238–9n9
- Emily (WASP, intersex, lesbian; Kitchener): on delayed elimination, 132; on gender-segregated bathrooms/toilets, 56, 60, 74; on hand washing, 138; on menstrual-product ads, 148; on mirror avoidance, 100; on speech in toilets, 114, 115; on urinary positions, 130; on white toilets, 86
- Eric Prete (mixed Ojibway and Celtic, trans man, bisexual; northern Ontario): on assaults, 79; on graffiti, 189; on passing, 58; on

peeing, 122; on urinary positions, 130; on voice for access to toilets, 119
- Farah (Caribbean-Canadian, transgender, queer; Toronto): on misreadings of predators, 76
- Frieda (British, non-trans, queer; Toronto): on defecating in public washrooms, 162; on gender-neutral toilets, 212
- Gypsey (Irish-Swedish and Scottish, transgenderist, lesbian; disabled; Toronto and Vancouver): on bathroom encounters, 56–7; on being considered 'dirty,' 156; on gender policing, 73; on smell, 152, 154; on the sounds of peeing, 126; on speech in toilets, 114; on urinary positions, 130; on washroom safety, 79
- Haley (non-trans queer femme): on defecating in public washrooms, 162; on the ideal bathroom, 257–8n2; on looks in 'men's' toilets, 91; on sex in public, 200; on signs on toilet doors, 214
- Isaac (white, Jewish, trans man, queer; Toronto): on being watched, 94; on cissexuals feeling tricked, 69; on expressions of hate as gendered, 77; on graffiti, 189; on peeing, 122; on washroom line-ups, 104
- Ivan (Portuguese-Canadian, non-trans man, queer; Toronto): on bathroom encounters, 55, 56; on choosing a urinal, 180; on code against touching, 186; on defecating in public washrooms, 165; on faeces and sex, 165; on gender-neutral toilets, 212; on graffiti, 188; on heterosexual washroom space, 169; on homophobic fear of attack, 193–4; on looks in 'men's' toilets, 92; on 'men's' room protocol, 180–1; on mirrors in bathrooms, 253n10; on perceived threat to children, 192–3; on policing public sex, 196, 197, 198; on public sex, 178, 198–9; on signs on toilet doors, 214
- Jacob (genderqueer): on defecating in public washrooms, 162; on signs on toilet doors, 214
- Jacq (British, genderqueer, butch, lesbian; disabled; Vancouver): on bathroom hostility, 57; on delayed elimination, 132; on expressions of hate as gendered, 77; on misreadings as predators, 75; on reversibility in visual exchanges, 65; on smells, 152, 154; on using breasts to access 'ladies' room, 66; on using the 'men's' room, 95; on washroom looks, 100–1
- Jannie (white, non-trans queer woman; Toronto): on feet in stalls, 94
- Jay (Irish and Italian, genderqueer, gay; Albany): on gender recognition in bathrooms, 59; on perceived threat to children, 192; on policing of public sex, 196; on purification rituals, 158; on sex in public, 201, 203
- JB (genderqueer, butch, lesbian): on gender, sexuality, and excretion links, 174; on graffiti, 189; on the ideal bathroom, 215–16, 257n2; on reversibility in visual exchanges,

65; on signs on toilet doors, 214; on urinals, 219
- Jersey Star (white, European, non-trans queer woman, polyamorous; Toronto): on public gay sex culture, 181; on sex in public, 201–2, 203–4; on smells, 152–3
- JM (genderqueer): on use of stalls, 93
- Joe (white, British, non-trans gay male; Montreal): on sex in public, 201; on sex in queer spaces, 200
- John (white, non-trans gay male; NYC): on graffiti, 186
- Kew (Chinese-Canadian, genderqueer, dyke): on ads in toilets, 254n11; on gender in queer spaces, 200–1; on the ideal bathroom, 216; on intersection of racism and genderism, 61; on policing public sex, 196; on sex in public, 200; on urinals, 219
- KJ (African-Caribbean, trans man, queer; Toronto): on leaving stalls, 95; on looks in 'men's' toilets, 91; on menstrual products, 150; on non-trans men policing gender, 72; on remains in the toilet, 143; on security guards, 70
- Lana (southern white, non-trans queer femme, gimp): on menstrual products, 150; on people with physical disabilities, 101, 102–3; on the sounds of peeing, 126; on urinary positions, 130; on using the 'men's' room, 95; on voice for access to toilets, 120
- Laura (Italian and Ukrainian, cis-gendered non-trans female, queer; Toronto): on ads in toilets, 254n12; on sex in public, 201
- Layal (Arab, non-trans queer femme; Toronto): on public as dirty, 159; on sex in public, 203
- Liam (white, non-trans gay male; Toronto): on policing public sex, 196
- Lisa (white, non-trans femme lesbian; Kingston, ON): on graffiti, 189
- Madison (Italian-American, genderqueer; New Jersey): on being considered unclean, 156; on gender-neutral toilets, 212; on graffiti, 187, 189; on the ideal bathroom, 215; on people with physical disabilities, 101–2; on sounds in toilets, 116; on touch in toilets, 157; on transphobic looks using mirrors, 88; on voice for access to toilets, 120–1; on washroom design, 85
- Meredith (Shanghai and Hong Kong Chinese, non-trans woman, queer; Toronto): on being considered less clean, 156
- Mykel (white, Anglo-Saxon, non-trans gay male): on looks in 'men's' toilets, 91
- Nandita (Pakistani, non-trans woman, bisexual; Toronto): on looks in 'men's' toilets, 91–2; on purification rituals, 158; on signs on toilet doors, 214; on washrooms as places of violence, 239–40n11
- Neil (Middle-Eastern, East Indian, trans guy, queer; southwest Ontario): on choosing toilets, 63;

on gender, sexuality, and excretion links, 174, 176; on looks in 'men's' toilets, 91; on racialization of toilets, 61; on sex in public, 204; on sociality in toilets, 115; on speech in toilets, 114–15; on voice for access to toilets, 121; on washroom looks, 89; on white in toilets, 86
- Pam (white, American, masculine/butch woman; Kingston, ON): on graffiti, 189; on sex in public, 200
- Phoebe (white, trans woman; Toronto): on people with physical disabilities, 101; on security guards, 70; on signs on toilet doors, 214; on voice for access to toilets, 120; on 'women's' washroom, 59, 142
- Property (Caucasian, queer female; Toronto): on misreadings of predators, 75; on peeing, 122
- Rachel (queer, butch): on gender-identification as mirrorical, 65; on hostility in bathrooms, 55; on using breasts to access 'ladies' room, 66
- Rachel (transgender, lesbian): on choosing a urinal, 180
- Raj (Pakistani, Muslim, non-trans gay man; Toronto): on sex in public, 199–200
- Rico (WASP, non-trans male, gay): on gender policing, 72; on graffiti, 188; on homophobic fear of exposure, 185; on masculine/feminine smells, 153; on purification rituals, 158; on sex in public, 201; on touching in toilets, 157; on using the 'men's' room, 95
- Rocky (visible cultural minority, genderqueer, non-trans female, queer; Buffalo): on the ideal bathroom, 216; on peeing, 122–3; on security guards, 72; on signs on toilet doors, 215
- Rohan (Anglo-Celtic, trans, butch, queer; Toronto): on ads in toilets, 148; on apologies by cissexuals, 67; on assault by non-trans man, 73; on choosing toilets, 63; on cleanliness, 245n1; on defecating in public washrooms, 162, 163; on defecation, 166; on disgust, 155; on erotic looks, 89; on gender-inclusive washrooms, 74; on graffiti, 189; on hostility in bathrooms, 55; on the ideal bathroom, 215; on menstrual products, 149; on mirrors, 88, 100; on misreadings of predators, 76; on perceived threat to children, 192; on reversibility in visual exchange, 65; on rituals to avoid touch, 157; on sociality in toilets, 115; on sound in toilets, 116–17; on using breasts to access 'ladies' room, 67; on voice for access to toilets, 119; on washroom design, 85
- Roxie (visible minority, trans female, bisexual; San Francisco): on homophobic fear of attack, 194; on perceived threat to children, 191
- s. applebutters (non-trans queer man): on being considered 'dirty,' 156
- Sarah (transgenderist; Ontario): on abuse and readings of sexuality, 77; on cleanliness, 137–8, 159; on delayed elimination, 132; on erotic looks, 89; on gender, sexuality, and excretion links, 174; on homophobic fear of attack, 194–5;

on perceived threat to children, 191; on security guards and police, 70, 72; on washroom line-ups, 103
- Sasha (Norwegian, genderqueer, bisexual; Chapel Hill, NC): on cissexuals feeling tricked, 69; on signs on toilet doors, 214–15
- Savoy (non-trans queer femme): on gender, 205
- Seo Cwen (U.K., non-operative trans woman in transition, gay; Kanata, ON): on homophobia in toilets, 180; on homophobic fear of attack, 194; on peeing, 122; on perceived threat to children, 191; on using gender-segregated toilets, 58, 59–60
- Shane (white, trans male, straight; San Francisco): on cissexuals feeling tricked, 69
- Shani Heckman (white, non-trans genderqueer, butch; San Francisco): on appearing gross, 156
- Skyler (British-Polish-Jewish, non-trans gay man): on policing public sex, 197–8
- Sugar (Anglo-Scottish and Celtic, non-trans queer femme; Vancouver): on ads in toilets, 254n11; on colour of toilets, 139; on speech in toilets, 115; on urinary positions, 130–1; on using stalls, 93
- Syd (Japanese, western-European, boi/fag, drag-queen bitch, queer, polyamorous; San Francisco): on graffiti, 190; on the ideal bathroom, 216
- Tall guy (white, non-trans gay male): on policing public sex, 197
- Tara (white, genderqueer, gay): on bathroom cleanliness, 137–8; on colour of toilets, 139; on remains in the toilet, 143; on urinary positions, 129–30; on white toilets, 86
- Temperance (WASP, non-trans queer femme; Toronto): on choosing toilets, 63; on delayed elimination, 133; on denied access, 53; on gender-segregated bathrooms, 55; on menstrual noise, 127; on public sex, 169; on the sounds of peeing, 126; on using stalls, 93–4; on washroom line-ups, 103
- Thomas Crown (French and Belgian, non-trans genderqueer man, gay): on gender, sexuality, and excretion links, 174; on homophobic fear of exposure, 185; on public sex, 200, 201
- Tom (bisexual): on looks in 'men's' toilets, 92
- Tulip (Israeli, non-trans genderqueer femme, bisexual; Brooklyn): on gender-segregated toilets, 63; on masculine/feminine smells, 154; on position of toilet seat, 143–4; on remains in the toilet, 143
- Velvet Steel (Danish, post-operative transsexual woman, bisexual; Vancouver): on defecation, 166; on delayed elimination, 132; on mirrors, 88; on panoptic designs, 85; on purification rituals, 158; on using gender-segregated toilets, 60
- Zac (white, English, non-trans gay male; Toronto): on homophobic fear of exposure, 186
- Zahabia (Indian-Canadian, Muslim, non-trans, queer, femme,

lesbian): on children in bathrooms, 57
– Zahara Ahmad: on ads in toilets, 148
– Zoe (non-trans woman; Ontario): on cleanliness, 245n3; on graffiti, 187–8; on the ideal bathroom, 215; on pornography in toilets, 179; on smells, 154–5
invagination (Derrida), 125
Irigaray, Luce, 5, 82, 112, 146–7, 160, 217–18
Isaksen, Lise Widding, 39–40

Juang, Richard, 4

Kipnis, Laura, 34–5, 203
Kira, Alexander, 7, 21, 82, 122
Kitchen, Rob, 80
Klein, Melanie, 13, 141
Kovel, Joel, 140
Krafft-Ebing, Richard von, 31
Kristeva, Julia, 127, 137, 141, 146, 166, 244nn12–13, 248n17

Lacan, Jacques, 32, 47–9, 84, 110, 233n5, 237n4. *See also* visual imago
Lahiji, Nadir, 6, 82, 135, 209, 210
Lambton, Lucinda, 31
Lane, Christopher, 178
language: about menstrual products, 148–9; of elimination, 21; gender pronouns and identity, 249n20; metonymic relationships in, 6–7, 41, 81, 252n4; of Parisian sanitary reform, 37; relationship with materiality, 54; rhetoric of racial slurs, 231–2n10

Laporte, Dominique *(The History of Shit)*, 32, 40, 154, 235n15
Law, Robin, 33, 80
law and legislation: activism challenging, 227–8n2; cases of denied access, 8–9, 175, 228–9n3; criminalization of homosexuality, 140; governing gender segregation in toilets, 70; human rights lawsuits, 73; policing sex in toilets, 195–204, 197; used in schools, 230n5
Leach, Neil, 106
Lefebvre, Henri, 84, 128
lesbian phallus, 46
Levinas, Emmanuel, 161
lighting in public washrooms, 85–6
Linde, Ulf, 240n1
literature: attention to the lavatory, 33–4; connecting elimination and sexual practices, 31; faecal imagery in, 248n17; sexuality studies, 18–19, 27; on toiletry practices, 174–5
London (U.K.): origins of gender-segregated toilets, 28, 134–5; regulating sex trade, 37–8; street excretion, 39; vertical and horizontal design of, 81
London Jilt, The, 37
London Science Museum, 31
Longhurst, Robyn, 4
Love, Brenda, 31

Magni, Sonia, 202
Malthus, Jane, 33
Marks, Laura, 136, 151, 159–61, 167
Marx, Karl, 167
masculine/feminine: and abject leaking bodily fluids, 146–50; acoustic

recognition of, 111–12, 114–21; as contiguous, 145; and defecating in public washrooms, 162–3; in dominant safety narrative, 74–5; effeminate men, 16; feminization of Asian men, 182–3; and forms of embodiment, 247n13; in the ideal bathroom, 217–18; and inhibiting touch, 166–7; interiors and exteriors confused, 118, 123–4; and the maternal, 117–18, 124, 127; and menstruation, 126–7, 163; and phantom phallus, 121–4; stink as masculine, 153–4; and trauma of gender, 213–14; and the urinal, 219–20

Masquelier, Adeline, 246n4

McClintock, Anne, 137, 140, 141

menstruation, 126–7, 146–51, 163

Merleau-Ponty, Maurice, 64, 65, 238n8

methodology, 13–24; of interviews, 20–4; terminology, 16–21

Metropolitan Transportation Authority (New York), 229n3

Mexico, 231n9

Miller, William, 128, 244n13

mimesis, 167

Minnesota Department of Human Rights, 229n5

Minor, Horace, 33

mirrorical returns, 82–4, 87, 99–100, 104, 206–7, 240n1

mirroring in transphobia and homophobia, 63–6

mirrors: bodies on display, 84, 113; and female urinals, 220; looks become erotic, 89–90; maintaining distance from, 99; and male cissexuals, 90–1; masculine and feminine in, 108–9; in 'men's' toilets, 90; and misrecognitions, 221, 237n4; to monitor sameness and difference, 49–50; not a neutral glass, 241n7; in panoptic design, 81; penis as hyper-visible in, 92 (*see also under* penis); race and public, 86; resistance and redeployment, 87–8; in reversibility in visual exchange, 65–6; sink and toilet bowl reflecting gender, 144. *See also* acoustic mirror; optical strategies; panopticism

misogyny, 188

misrecognitions, 47, 48, 50; all recognitions as, 221, 237n4. *See also* gender recognition (achieving)

modernity, 32, 34, 173, 245n15

Molesworth, Helen, 82, 148–9

Moore, Alison, 36

Morgan, Margaret, 34, 236n16

Muñoz, José Esteban, 49

museum, lavatory as, 41–2

Museum of Sex (New York), 31

mysophilia, 136

Namaste, Viviane K., 18

National Student GenderBlind Campaign (U.S.), 227n2

New York City: graffiti-based bathroom installation in, 221; Grand Central Station, 175, 228–9n3; sewer system, 233n2. *See also individual interviewees*

Noble, Bobby, 249n20

noise. *See* acoustic sensations

Oedipal complex, 29, 35, 47, 121, 123, 128, 210, 213. *See also* Freud, Sigmund

Office and Technical Employees Union (BC), 9
olfactory sensations. *See* smell and odour
Olympic sex testing, 43
Ontario Human Rights Code, 9, 70
optical strategies: haptic visuality, 159–61; in 'men's' toilet, 90–1; in 'women's' washroom, 87–90. *See also* mirrorical returns; mirrors; visual imago
orifices: desire centred on unauthorized, 172; disciplining of, 30; in ego formation, 47–8; as leaky, 146–7; the navel, 213; of women, 35; the word, 233n3

panopticism: acoustic, 105–6; development of, 79–81; historical use of, 5–6; and mirrors in washrooms, 87, 253n10; of stalls and urinals, 92–6; and ways of seeing, 81. *See also* architecture and design (of washrooms); mirrors; surveillance
Paris: origins of gender-segregated toilets, 28, 134–5; regulating sex trade, 37–8; sewer system, 233n2; street excretion, 39
parodic performances, 104, 111
paruresis, 241n5
passing, 58–9, 69, 238n7
Pears' Soap, 140
Peeing in Peace (Transgender Law Center, San Francisco), 227n2
penis: as hyper-visible, 92, 112–13, 185; and looks in 'men's' toilets, 91–2; and phantom phallus, 121–4; and racialized men, 182–3; and transphobia, 240n3. *See also* anal eroticism and elimination

Penner, Barbara, 31, 220
People in Search of Safe and Accessible Restrooms (PISSAR), 227–8n2
Pike, David, 41–2
Pile, Steve, 11–12, 121, 123, 125, 128, 249n22
plague, 42, 134. *See also* disease and dirt
plumb line, 206–8; and the epistemology of the plumber, 209–15
plumbed bodies, 152
policing of gender: by gays and lesbians, 13; historic context of, 6–7; laws governing segregation in toilets, 70; by non-trans men, 72, 92; people as excrementalized, 137; by police, 71; in public washrooms, 10, 69–73, 231n7; by security guards, 69–70, 72, 196
policing sex in toilets, 177–9, 195–204
pornography, 38, 179, 203
privacy in toiletry habits: heteronormativity of, 7–8; history of, 201; and loss of communal activities, 34–5; and olfactory intolerance, 245n2. *See also* public space
privacy rights, 4
projective identifications, 13, 157–8. *See also* disidentifications
Pronger, Brian, 163, 164
Prosser, Jay, 44, 86, 109, 170
prostitution or sex trade, 37–8, 244–5n14
psychoanalysis, 43, 44, 96–7. *See also* bodily ego and gender; Freud, Sigmund
public health and gender-segregated toilets, 135, 151
public sex: about public-private distinctions, 198–204; and closing of

public toilets, 253n7; gay male public sex cultures, 181–2; prohibitions about gender, 171–3; removal of doors and, 177–8, 253n9. *See also* policing sex in toilets
public space: accessibility and safety of, 8; determined by heterosexual self-identifications, 199; humane recognitions and participation in, 62 (*see also* humane recognitions); made to feel like home, 217; and production of gender, 12–13; as rigidly gendered, 53; sex as belonging in, 198–204; urinal as public and private, 219. *See also* privacy in toiletry habits
public washrooms: colour in, 86, 139–40 (*see also* white and whiteness); defecating in, 161–6 (*see also* acoustic sensations); and fantasies of home, 45 (*see also* home and heteronormativity); 'Gents' (South End Green), 139 (*see also* history and cultural politics of excretion); in interviews, 20–1; line-ups for, 103–4; and pedagogy of examination, 43; use of term, 21; 'women's' versus 'men's' rooms, 92, 114, 115, 137–8. *See also* gender-segregated toilets; gender signs on bathroom doors; sanitary technology; stalls; urinals, male; urinals and urinettes, female; violence
purity: as absence of smells, 152; architecturally mandated, 82; colonial ethic of gender, 140; and colour, 139; and disease and psychic loss, 41; established by abjection, 137; established by voice, 114; and sounds/silence, 106, 109; and toilet technology, 37. *See also* hygiene and hygienists

qualitative research data, 20. *See also* interviewees (methodology)
queer, use of term, 14, 19–20
queer sex in bathrooms. *See* public sex
queer theory, 18–19, 232n18. *See also* literature

'race' and racialization: and colour of bathrooms, 86, 139; in design of washrooms, 13, 61, 251n30; disciplinary institutions creating, 86; and ethic of gender purity, 140; of gay male public sex cultures, 182–3, 254–5n13; and gender purity, 7; in graffiti, 189; and hand washing, 246n4; imagery and acoustics, 120–1; and incongruity in bathrooms, 49; and reproduction, 255nn14–15; and rhetoric of faecal matter, 231–2n10; of the vagina, 125; water closet and, 11, 233–4n7. *See also* white and whiteness
Rawes, Peg, 5
Reddy, Vasu, 202
religion. *See* Christianity
reproduction, 202, 255nn14–15; and unproductive sex, 252n2; white reproductive futurism, 57. *See also* family (heteronormative)
ritual of urination, 129
Rose, Gillian, 12
Rosolato, Guy, 118
Rush, Fred, 123, 218

Sacco, Lynn, 248n16
Sade, Marquis de, 36, 248n14

safety: cases demanding, 8–9, 229–30n5, 231n7; construction of masculinity in narratives of, 74–5; and feminine hygiene, 150–1; and the ideal bathroom, 215–16; lacking in gender-segregated toilets, 73–8; narratives obscuring patterns of violence, 75; and perceived threat to children, 192; as rationale for gender-segregated toilets, 63, 135; threat of homosexual or trans subject, 176–7. *See also* delayed elimination; violence

Salamon, Gayle, 45–6, 48, 97

San Francisco. *See individual* interviewees

sanitary engineers and city planners: regulating sex trade, 37–8, 42; in story of elimination, 32–3; and threat of disease, 40. *See also* hygiene and hygienists

sanitary technology: no-touch washroom fixtures, 167–8, 251n30; and surveillance, 79

'scatontological anxiety,' 51

Schilder, Paul, 47, 98

school washrooms, 8, 229–30n5

science: comparisons in design of washrooms, 85; and hygienic superego, 145; no rationale for gender-segregated toilets, 63; and pedagogy of examination, 42–3; as a religion to manage bodies, 34; systems of regulation, 6. *See also* bio-political regulation of the body

Sedgwick, Eve Kosofsky, 62

sensory systems: haptic visuality, 159–61; hierarchy in design, 136, 141, 151; relation with self-image, 98–9; use of rhythms and colours, 123. *See also* acoustic sensations; smell and odour

Serano, Julia, 8, 16, 53–4, 58; on passing, 238n7

sewer systems, 7; and class, 233n4; compared to sex organs, 37–8, 42; drowning in, 41, 236n18; and the epistemology of the plumber, 209–15; plumb line, 206–8; and regulating sex trade, 37–8; sanitizing the public, 42; as underworld, 136, 233n2. *See also* history and cultural politics of excretion

SexGen committee (York University), 228n2

sex-segregated toilets. *See* gender-segregated toilets

sexual difference: and body boundaries, 246–7n11; and defecation, 164–5; encountering gender difference, 55; gender neutral washrooms as threat to, 5; loss of binary gender axis, 4; managed by inhibiting touch, 166–7; in origins of gender-segregated toilets, 28; signifiers of, 10, 113–14; space in regulation of, 10–13; in urinary capacities, 30. *See also* urinary positions

sexuality studies, 18–19, 27. *See also* literature

shame, 60–2

Shildrick, Margrit, 49

shit and anal imaginations, 161–6. *See also* anal eroticism and elimination

Sibley, David, 12

Siemens, 9

signs on bathroom doors. *See* gender signs on bathroom doors

Silverman, Kaja, 48, 49, 50–1, 83; acoustic mirrors, 107–9, 243n8; on bodily ego, 97–100, 172, 241n6; castration, 242n3; de-idealizing gaze, 87; on female and male voice in film, 111, 117, 242n7; on feminine (maternal) speech, 117–18, 241–2n1; on femininity, 123; identifications at a distance, 68; on masculine space, 125; on the sensational body, 146; on sounds marking the feminine, 127; vagina and the maternal voice, 243n11; on voices and female genitalia, 119
Silverman, Michael, 175
sinthomosexual, 177, 257n22
slums. *See* class
smell and odour: and class, 11–12; and haptic visuality, 159–61; heteronormative management of, 152–3; learned response to, 151–2, 154; and the mechanics of disgust, 151–6, 160, 245n2; policing of, 6; purity as absence of, 152
sodomy, 10, 135, 140, 165–6, 183–4, 201–2
space. *See* architecture and design (of washrooms); public space
Spade, Dean, 228n3
stalls: as confessionals, 256n17; door removal, 177–8, 243n10, 253n9; as feminine, 112–13, 123; in the ideal bathroom, 215; panoptic design of, 92–6, 253n6; promiscuous looks in, 94; talk between, 115–16. *See also* toilet bowl; urinals, male; urinals and urinettes, female
Stallybrass, Peter, 43, 81
stink. *See* smell and odour
Stoler, Ann Laura, 86

Stone, Helena, 175, 228–9n3
Stryker, Susan, 17, 19, 54, 62, 100
students and school washrooms, 229–30n5
subject dissolution. *See* identity and identity categories
surveillance: acoustic, 105–6; in design of public washrooms, 82–4; door removal, 177–8, 253n9; gender under the microscope, 84–92; of 'men's' room, 195; and pedagogy of examination, 42–3; public toilets in travel hubs, 52; rise of, 79; of stalls and urinals, 93–6. *See also* panopticism
Sylvia Rivera Law Project (New York), 227n2

taboos, 29, 30, 199–200
taste and haptic visuality, 159–61
teachers, 229–30n5
Thomas, Calvin, 51, 118, 137
toilet bowl: as cloistered, 113; and hand washing, 138; remains on, 143–4, 157; and urinary positions, 81, 129. *See also* architecture and design (of washrooms); stalls; urinals, male; urinals and urinettes, female
toilet humour, 248n15
toilet lid and seat, 143–4
toilet paper, 36–7, 138
toilets: as abject, 141; various words for, 21. *See also* public washrooms
toilet technologies: class divisions and, 234n9
toilet training, 35–6, 129, 211, 213, 250n23. *See also* children
Toilet Training: Law and Order in the Bathroom (film), 227n2

Toronto: gay male public sex culture in, 181–2 (*see also* interviewees); interviews and interviewees, 20; malls, 194, 196; police surveillance, 178–9, 257n25; street excretion by men, 39, 236n17; Timothy Eaton, 28; transit system, 196. *See also individual* interviewees
touch and touching: disrupting codes against, 186; and haptic visuality, 159–61; intimacy and effect of, 156–7; and managing sexual difference, 166–7; and no-touch fixtures, 167–8; policing of, 6; rituals to avoid, 157–9
tourism: and excretion, 33
tranny-queer: use of term, 231n8
trans, 16. *See also* transgender, transsexual, and/or genderqueer
trans bodies: regulation of, 6
transgender, transsexual, and/or genderqueer: identity categories of, 14; status as castrated subjects, 109; transgender rage, 62; transgender studies, 17–19; use of terminology, 16, 17–20; visual imago and sensational body for, 98–9; and women's construction of masculinity, 74. *See also* bodily ego and gender
Transgender Law Center (San Francisco), 227n2
transgenderist people, 16; the term, 238n5
transitions, negotiations of, 44
transphobia and homophobia: among LGBTI people, 18–19, 23; and anxiety over defecating, 162; in broken mirror circuits, 65–6; and call for gender-neutral toilets, 211–12; codes of conduct, 185; disgust as response of, 155–6; distinctions, 10, 15; fear-based responses as, 76; and focus on children, 190–3; gender-normative imagery, 64–6; in graffiti, 186–90; looks become erotic, 89–90; and mirror avoidance, 99; and narrative gay/straight attack, 193–5; securing signifiers of difference, 113–14, 256n21; shame effect of, 61; and 'shit out of place,' 163–4; and surveillance in washrooms, 240n3; and toilet spatial configurations, 180; use of mirrors, 88; using breasts to access 'ladies' room, 66–7; voyeurism, 178–9. *See also* violence
trauma as identification, 213
travelling, 52
Trotter, David, 110
two-spirited people, 16, 60–1, 71, 132, 217, 219. *See also* transgender, transsexual, and/or genderqueer
Twyford, Thomas, 243n9

underground, 38, 41–2, 136. *See also* sewer systems
uprightness, 207–8, 219–20
urban space, 12, 20, 38. *See also* London; New York City; Paris; Toronto
urinals, male: amplifying male genitalia, 112–13; design resembles vagina, 122; versus enclosed oval toilet bowl, 11; and hand washing, 138; history of, 28; and the ideal bathroom, 218–19; 'mirrorical return,' 82–3; panopticism of, 92–6; as public and private, 219; as used by gay males, 131; as used by transmasculine folk, 129–33, 240n3. *See also* choreography of the body; stalls; urinary positions

urinals and urinettes, female, 28, 31, 39, 220–1; choreography of the body, 80. *See also* stalls; toilet bowl
urinary positions, 81, 128–33, 183–5, 244n14
urine as ungendered, 143

vagina and clitoris: associated with the maternal voice, 243n11; urinal design and, 122; as visible/invisible, 112–13. *See also* genitals
Valentine, David, 14, 19, 73
Vasseleu, Cathryn, 65, 238n8
vice. *See* prostitution or sex trade
Victorian era: colour in washrooms, 139; covering sounds of elimination in, 243n9; hygiene and religion, 34; preoccupation with the body, 27–8, 31–2; Queen Victoria, 36–7, 243n9; segregated public space of, 38–9; sexuality and the toilet in, 36. *See also* history and cultural politics of excretion; water closet
violence: attacks in public washrooms, 9–10, 229n4, 239–40n11; beaten in the toilet, 220; and fear in gender-segregated toilets, 41, 60, 69–73; forceful removal, 62; and masculinity constructions, 77; narratives as obscuring, 73–4; by non-trans men policing gender, 72–3; or living with incoherence, 78; patterns obscured in safety narratives, 75; and perceived threat to children, 193; psychic acts of, 54; result of gender panic, 76–7, 195–7. *See also* safety; transphobia and homophobia
Virgin/Whore dichotomy, 113

vision. *See* optical strategies; visual imago
visual imago: and alignment with sensational body, 48–9; gap with sensational body, 86, 98–100, 241n6; and public mirror circuits, 66

Wallon, Henri, 48
Warner, Michael, 173, 204
water closet: capitalism and, 234n8; colonialism and, 61; invention of, 6, 227n1; and modernity, 34; in production of whiteness, 11, 233–4n7. *See also* history and cultural politics of excretion; Victorian era
Weinberg, Martin S., 246–7n11, 247n13
Wesley, John, 34
White, Allon, 43, 81
white and whiteness: about abstinence and absence, 145–8, 166–7; in bathroom design, 246n9; and death, 246n5; and menstrual blood, 147–8; porcelain, 11; produced in water closet, 233–4n7; as signifying, 139; use of colour in washrooms, 139–40; water closet and production of, 11; and white in toilets, 86. *See also* 'race' and racialization
Whittle, Stephen, 16
Williams, Colin J., 246–7n11, 247n13
Wilton, R.D., 12, 50
workplace toilet access, 230n6
Wright, Lawrence, 38

York University (Toronto): gender-neutral lavatories, 228n2
Young-Bruehl, Elisabeth, 192, 256n21

Žižek, Slavoj, 40